Has It Come to This?

Nature, Society, and Culture

Scott Frickel, Series Editor

A sophisticated and wide-ranging sociological literature analyzing nature-society-culture interactions has blossomed in recent decades. This book series provides a platform for showcasing the best of that scholarship: carefully crafted empirical studies of socioenvironmental change and the effects such change has on ecosystems, social institutions, historical processes, and cultural practices.

The series aims for topical and theoretical breadth. Anchored in sociological analyses of the environment, Nature, Society, and Culture is home to studies employing a range of disciplinary and interdisciplinary perspectives and investigating the pressing socioenvironmental questions of our time—from environmental inequality and risk, to the science and politics of climate change and serial disaster, to the environmental causes and consequences of urbanization and war making, and beyond.

Has It Come to This?

The Promises and Perils of Geoengineering on the Brink

EDITED BY J. P. SAPINSKI, HOLLY JEAN BUCK,
AND ANDREAS MALM

Rutgers University Press

New Brunswick, Camden, and Newark, New Jersey, and London

Library of Congress Cataloging-in-Publication Data

Names: Sapinski, J. P., editor. | Buck, Holly Jean, 1981– editor. | Malm, Andreas, 1977– editor.
Title: Has it come to this? : the promises and perils of geoengineering on the brink / edited by J. P.
Sapinski, Holly Jean Buck, and Andreas Malm.
Description: New Brunswick : Rutgers University Press, [2020] | Series: Nature, society, and culture |
Includes bibliographical references and index.
Identifiers: LCCN 2020005630 (print) | LCCN 2020005631 (ebook) | ISBN 9781978809352
(paperback) | ISBN 9781978809369 (cloth) | ISBN 9781978809376 (epub) | ISBN 9781978809383
(mobi) | ISBN 9781978809390 (pdf)
Subjects: LCSH: Environmental geotechnology. | Climate change mitigation.
Classification: LCC TD171.9 .H37 2020 (print) | LCC TD171.9 (ebook) | DDC 628.5/32—dc23
LC record available at https://lccn.loc.gov/2020005630
LC ebook record available at https://lccn.loc.gov/2020005631

A British Cataloging-in-Publication record for this book is available from the British Library.

♾ The paper used in this publication meets the requirements of the American National Standard
for Information Sciences—Permanence of Paper for Printed Library Materials, ANSI Z39.48-1992.

www.rutgersuniversitypress.org

Manufactured in the United States of America

Contents

Has It Come to This?

Part I

Introduction

1

Critical Perspectives on Geoengineering

A Dialogue

HOLLY JEAN BUCK, J. P. SAPINSKI,

AND ANDREAS MALM

"Geoengineering is inevitable," claimed a recent headline in *Earther*, predicting a raft of policy reports and think pieces that will increase in volume and
urgency.[1] Indeed, every year around the time of the U.N. Framework Convention on Climate Change (UNFCCC) conference, such articles appear. The
Intergovernmental Panel on Climate Change's (IPCC's) special report *Global
Warming of 1.5°C*, released in October 2018, indeed delivered a dire message: we
have a decade to slash carbon dioxide emissions by 60 percent before breaching
the 1.5°C target.[2] Even resilient and transformation-minded earth system scientists like Johan Rockström predict that such findings "will raise solar radiation
management to the highest political level," as he told the *Guardian* when the
report was published, adding, "I'm very scared of this technology but we need
to turn every stone now."[3]

How did we get into this situation? Is it even an accurate representation of
where we are at? Who is "we," anyway? What else could be done, and what
options make it onto the table? Which are left out? Whom does geoengineering serve? Why is the ensemble of projects that goes by that name so salient
even though, as Ina Möller points out in chapter 2, the community of researchers and advocates is remarkably small? These are some of the questions that
the thinkers in this volume are exploring, drawing from fields such as human
ecology, geography, sociology, science and technology studies, ethics, political
economy, and Indigenous studies. We set out this diverse collection of voices
not as a monolithic, unified take on geoengineering but as a place where creative thinkers, students, and interested environmental and social justice advocates could explore nuanced ideas in more than 240 characters. In a world full
of "hot takes" on a hot climate, filled with urgency and fear—and the climate
crisis is truly frightening—we wanted to slow down and develop a critical analysis, linking this emergent issue to theory, experience, and insights from other
fields. The thinkers here have spent time on their analyses, producing a scholarly

complement to work being done by social movements on the ground. We invite you to question geoengineering approaches to climate change with us.

"Geoengineering" in the Era of Overshoot

In 1977, Cesare Marchetti, based at the International Institute for Applied Systems Analysis in Austria, wrote the back-of-the-envelope paper "On Geoengineering and the CO_2 Problem." This thought experiment treated the climate problem as one of "global kinetics" and looked at collecting, transporting, and disposing of CO_2 in the deep ocean (in a system dubbed the "Gigamixer").[4] Marchetti's may have been the first use of "geoengineering" as a signifier, though ideas of weather modification or fantasies of climate control stretch back further still.[5] Coincidentally, 1977 also saw the translation of Russian climatologist Mikhail Budyko's book *Climatic Changes*,[6] which brought to an English-speaking audience the idea of blocking incoming sunlight to cool the earth by spreading sulfur particles in the stratosphere. As Möller chronicles in chapter 2, the groundwork for geoengineering as a concept was laid in the early 1990s. The early 2000s saw a renewed interest in Budyko's approach, which came to be called "solar geoengineering" or, in a peculiar euphemism, "solar radiation management." It was also at this time that modelers began to incorporate "overshoots" of temperature and greenhouse gas (GHG) emissions targets into their modeling, as Wim Carton discusses in chapter 3. This naturally brought up the question of carbon capture and storage (CCS). As Chris Huntingford and Jason Lowe wrote in a 2007 letter in *Science*, responding to an article on "CO_2 Arithmetic," "If emissions targets are not met, or if the impacts of climate change are greater than expected, we might well find ourselves in the position of having a greenhouse gas level that is 'dangerous,' thus accidentally following an overshoot scenario. In this case, the capability to draw CO_2 from the atmosphere might be highly desirable."[7]

In 2009, the British Royal Society published a seminal report called *Geoengineering the Climate*.[8] It threw up two ideas that were reproduced countless times over the next decade. The first was the idea of "geoengineering" as a two-part family: solar radiation management (SRM) on one side and carbon dioxide removal (CDR) on the other. To make these categories intelligible to the noninitiated—and so far, geoengineering has very much been a topic reserved for specialists—we might break it down into the basics. SRM is about turning down the heat on Earth by reducing the amount of incoming sunlight. This can be achieved in several ways. As detailed in the Royal Society report, strategies in the SRM basket include space reflectors (structures that would reflect sunlight from outer space—think a fleet of mirrors orbiting around Earth), cloud albedo modification (whitening clouds over the oceans so that their albedo—their ability to reflect sunlight back into space—is increased; special ships might spray seawater into the air to brighten the clouds), and modification of the surface albedo (spreading plastic sheeting in deserts or on melting glaciers or designing whiter surfaces in urban areas—think painting roofs white). The most salient SRM scheme, however, is known as stratospheric aerosol injection. Here the idea is to spread sulfur or other particles into the stratosphere so as to create a sunshade.

The second category, CDR, attacks the climate equation from another direction. It seeks to pull down carbon dioxide from the atmosphere and store it somewhere on Earth. This would attenuate the greenhouse effect, and again, the strategies conjured up for attaining this goal are multiple. The Royal Society listed biochar (a kind of charcoal that is mixed into soil so that it stores carbon), capture and sequestration of carbon dioxide at power plants (think a filter inside chimneys that scrubs the CO_2 molecules from the smoke and leads them out so they can be piped and buried under the ground), enhanced mineral weathering (adding minerals to accelerate the natural process by which rocks dissolve in rainwater and suck up CO_2 from the air), afforestation (growing forests [trees swallow CO_2]), and ocean fertilization (adding nutrients to the ocean to feed plankton; more plankton means more CO_2 drawn into the ocean; upon death, some of the extra plankton would drizzle down into the seabed and take the CO_2 with them). One instrument in the CDR toolbox has received the most attention in models: bioenergy with carbon capture and storage (a.k.a. BECCS). Here the idea is to establish extensive plantations of eucalyptus or some other fast-growing plant, have them drink CO_2 from the air through photosynthesis, harvest them, and send them off to power plants, where they are burned; in the process, energy is generated while the carbon is captured through a filter and then injected, in concentrated form, into cavities under the ground. More recently, entrepreneurs have been working on direct air capture (DAC), which involves machines that capture carbon directly from the air to be injected underground in a similar way or to be transformed back into fuels.

The Royal Society's division of the geoengineering terrain into SRM and CDR stuck. The binary structure set up the conceptual landscape as a choice, and some of the early public engagement work treated it as such—publics were asked which they preferred, and they were found to prefer the more natural CDR over the more technological SRM, for example. The second innovation of the Royal Society report is the "preliminary overall evaluation of the geoengineering techniques,"[9] affectionately called the "blob diagram" by some scholars. This diagram, reproduced in figure 1, plots technologies along two axes—effectiveness and affordability—and then colors the technologies according to "safety" and sizes the dots according to "timeliness" (how quickly they could be implemented). The work this diagram performed was to determine the axes for assessment. It comes with the caveat that "this diagram is tentative and approximate and should be treated as no more than a preliminary and somewhat illustrative attempt at visualising the results of the sort of multi-criterion evaluation that is needed."[10] Nonetheless, perhaps predictably, it was in fact seized upon and used in countless PowerPoint presentations to come, which were unable to capture all the written caveats—that what it illustrates is that no ideal method exists, that the error bars should really be much longer but that this would confuse the diagram, and that it could "serve as a prototype for future analyses when more and better information becomes available."[11] Though subsequent publications suggested other criteria for assessment,[12] debates never moved very far from the terms set by the Royal Society.

However, this version of geoengineering laid out a decade ago—of colorful blobs representing speculative technologies spread out in a kind of

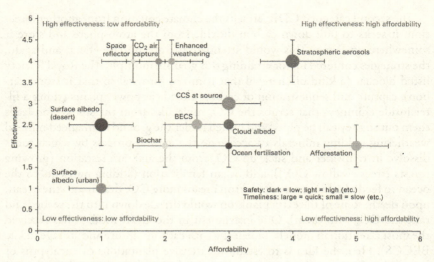

FIG. 1. The "blob diagram"

smorgasbord to pick from—is slipping away, and what's replacing it has a grimmer and grittier feel. The urgency of climate change is becoming ever clearer—as is the fact that integrated assessment models (IAMs) project that keeping warming below 1.5°C relies on CDR.[13]

Geoengineering's life as an umbrella term is also constantly under question. Today, instead of the Royal Society's SRM + CDR = geoengineering, "solar geoengineering" has become more familiar as a discrete approach. And carbon removal more often appears under newer names such as negative emissions (which in turn can be part of "deep decarbonization"), net-zero emissions, or even carbon capture and utilization. This book addresses both solar geoengineering and carbon removal, sometimes separately and sometimes together. This is because they are both global-scale, technology-informed approaches that, if deployed, will most likely appear in the context of an overshoot. Put differently, it might come to pass that targets for limiting global warming are breached and that some geoengineering option is rolled out to return Earth to safer levels—something akin to falling over the brink and then jumping back up by means of some ingenious technological contraption activated in the air. What could do the trick? We suggest that the idea of temporary solar geoengineering cannot be ignored: it is going to be an important part of the discourse even if it ultimately does not see the light of day. CDR, including DAC, can be even less discounted. On the table is also a portfolio response of mitigation, adaptation, carbon removal, and solar geoengineering—everything that can alleviate the catastrophe of climate breakdown.

For solar geoengineering, there are essentially just two end games. It could be enacted indefinitely, with particles continually placed into the stratosphere to maintain lower temperatures, possibly in ever-greater volumes to counteract continued CO_2 emissions, but this comes with a well-known risk. The sulfur particles would suppress the effects of excess CO_2 in the atmosphere, much like a lid kept with a firm hand over a boiling kettle. But what if the hand is lifted? What if the social or technical infrastructures of solar geoengineering fail or break down for any conceivable reason? Then the heat kept in check by the lid would rapidly boil over in what is technically known as a "termination shock." It could set the planet on fire. Species that might potentially adapt to more grad- ual rates of warming would be wiped out and ecosystems scorched. For this and many more reasons,[14] scientists and political analysts are both wary of indefinite solar geoengineering.

However, a second end game has been posited: the temporary scenario, which would "shave the peak" off the warming "overshoot." As part of a "portfo- lio" response to climate change, the idea is that solar geoengineering is gradually ramped up over some decades and then gradually phased out again after econo- mies have decarbonized and carbon removal technologies have been developed to lower GHG concentrations.[15] In a rational world, this scenario would allow more time to scale up the massive institutional and technical infrastructure sup- porting practices and technologies that could remove carbon from the atmo- sphere. It implies that humans can effectively intervene in the global carbon cycle, if not manage it, given a little bit more time to enact change. Call it a dystopian-utopian prospect.

Critical Analyses

Up to now, scholars critical of geoengineering have approached the topic from two angles. A first critical strategy focuses on deconstructing the discourse around geoengineering, either the media discourse or that found in scientific publications. For example, Jane Flegal and Aarti Gupta critique how expert advocates of geoengineering invoke narrowly framed equity questions in ways that invite further scientific research.[16] Ethicist Duncan McLaren[17] finds that public debate in the media revolves around frames of technological optimism, political realism, and catastrophism. Within this discourse, confidence in the capacity of technological strategies to address climate change combined with "realist" acceptance that negotiations have failed as the climate threshold is get- ting ever closer—if it hasn't been passed already—together work to depoliticize discussions and implicitly push for technoscientific governance of the field. Questions of justice are evacuated, and more radical political ecology and politi- cal economy framings that tend to reject geoengineering are marginalized. Mike Hulme's analysis of scientific debates finds similar framings that revolve around metaphors of planet Earth as a body or a patient in need of emergency treat- ment.[18] Critical of such formulations, historian James Fleming sets the historical use of this medicalized language within a broader world view that considers the biosphere as a system to be managed by humans who are in some way separated from it, which he argues is symbolically prior to geoengineering proposals.[19] Going further, Timothy Luke argues that geoengineering, as a technology of domination of nature, implies a reorganization of society that would bring in new modes of domination of humans, deep changes that are being obscured in the discourse of "experts."[20] Finally, bringing the focus to some of the social actors behind this discourse, Holly Jean Buck finds that in the few years following the Royal Society report, the media called upon a restricted group of scientists and experts who thereby got to frame the popular discourse on geoengineering during this crucial formative period.[21]

Overall, critical scholars thus find that the discourse around geoengineering comes from a very specific standpoint—that of a restricted number of scientists and engineers who view themselves as managers of the "earth system" in the face of the existential threat of climate change. In such discourse, political debates are sidelined in favor of a technological and technocratic approach that supports the economic status quo. Proposals of a more regulated capitalism, let alone more transformative approaches to transition to a different economic model altogether,[22] get dismissed as not "politically realistic."[23]

A second, closely related angle of critique considers the formidable ethical challenges posed by simply imagining such technologies. Several issues have been outlined that became part of the standard discussion of geoengineering ethics. For one, the fact that operational geoengineering technologies could be viewed as an alternative—a potentially cheaper one—to emissions reductions has been discussed as posing a "moral hazard": a temptation to pursue technofixes instead of actually cutting emissions. Some have argued that even research on geoengineering might undermine political negotiations around mitigation efforts.[24] CDR techniques and their inclusion in modeling scenarios have received particular attention here as just another flight from what needs to be done. Additionally, as governments and private actors invest large sums of money to create an institutional research structure, various interests become entrenched, making it harder to scale down commitments and creating strong pressures for moving toward implementation. This dynamic in which pressure from interested groups may outweigh the need for caution is referred to as "path dependency" or "technological lock-in."[25] This dynamic has been discussed as a "slippery slope," as geoengineering could become established before problems even emerge, and if problems do emerge, it would be too late to address them fully.[26] Thus the early framing of geoengineering as "plan B" in case emissions fail to be reduced in time poses some major problems: simply knowing that the possibility exists—the "promise" of geoengineering—sets forth a movement toward its implementation that then becomes institutionalized, making actual implementation more likely and possibly mitigation less so.[27] Here it would be a case of throwing oneself over the brink in the belief of a safety net working so well that no stepping back is needed.

Because of the deep concerns entailed, critical analysis of the geoengineering discourse and its political and ethical shortcomings has found its way into the broader literature on the topic. Nonetheless, such analyses approach the subject from within scientific discourse and institutions and can therefore only push the critique so far. Runaway climate change, to which geoengineering is imagined as a possible response, is the product of broader dynamics in the political economy of capitalism and the metabolic relationship between human society and the biosphere. Hence a third angle of critical analysis sets geoengineering in the social, political, and economic context in which the idea emerged, evolves, and will—perhaps—become reality. The fossil fuel economy unleashed by the industrial revolution provides the backdrop in which the climate crisis unfolds and is being addressed. This social structure is characterized by rampant social inequality, a symbiotic relationship between economic production and state power, and the entrenched power of the corporate elite. Understanding why such an extreme response to the climate crisis is now making its way into mainstream political thinking requires an assessment of how these social relationships constrain and enable responses to the climate crisis.

This book consolidates this emerging approach to geoengineering that takes into account the political-economic conditions within which it emerged. Capitalism as a mode of producing the goods and services used in everyday life is most often invisible to actors who are part of its functioning. This is also true for the vast majority of analyses of geoengineering: it is at best assumed that political-economic structures cannot be modified by means of social action, and capitalism is hence bracketed in the analysis; at worse, capitalism does not enter the picture at all. Whichever the case, a crucial element of the world in which geoengineering is considered is left out of the discussion. The authors of the following chapters build on a recently emerging literature to fill this major gap as they discuss the many issues raised by geoengineering by explicitly considering them as taking place within a capitalist economy, including inherent issues of gender, racial, and class inequality; environmental justice; state action; and corporate power. The book picks up the threads of the early analysis laid out previously and brings them together to address the following questions: Why geoengineering? Why all the buzz about it here and now? Under what conditions might geoengineering be pursued in a capitalist context? In what direction would geoengineering take capitalism if implemented?

Road Map for This Volume

This first chapter is dedicated to providing a road map for this volume. The remainder of the book is divided into three parts that work together to unpack different forms of oppression and injustice that geoengineering might interact with and articulate them with issues of state power, North-South relations, and imperialist expansion. The second part explores how social and environmental justice, Indigenous, and intersectional perspectives can be mobilized to challenge geoengineering from the perspective of civil society. But first, why did geoengineering rise to prominence in the global agenda despite the idea's obvious drawbacks? This question is asked by Möller, whose contribution approaches the concept of geoengineering with an ethnographically informed look at its institutional roots in chapter 2. Möller takes a look at the conditions that allowed geoengineering as an umbrella concept to become prominent. She describes the processes of rendering geoengineering a governable object as well as the network effects that allowed knowledge brokers to spread the idea of geoengineering into new settings. The chapter analyzes the sudden spread of geoengineering as a policy option—its origins within a small community of scientific experts, its diffusion through an increasingly well-connected knowledge network, the scientific context in which the idea was able to resonate, and the targeted diffusion to realms beyond science and the Western world. In the process, it discusses the importance of scientific authority, the function of demarcation and categorization in shaping our reality, and the role of policy entrepreneurship in promoting particular solutions. The chapter concludes with a reflection on how understanding the origins of a policy option like geoengineering can help us make more thoughtful decisions about how to respond to it.

Carton's chapter accompanies Möller's in looking into the roots of these approaches, as it focuses on negative emissions and the new reality they have helped perform since their introduction in IPCC modeling scenarios. The inclusion of speculative negative emissions technologies (NETs) in climate scenarios allows the IPCC or companies like Shell to construct visions of the future in which fossil fuels continue to be the main source of energy far into the future. Carton tempers the hopes world governments are putting in what he calls a "convenient fiction," outlining the many shortcomings and current technical hurdles of NETs. His chapter argues that reliance on such speculative strategies is at best utopian and leads us astray from proven strategies to reduce GHG emissions.

Chapter 4, authored by environmental campaigners Lili Fuhr and Linda Schneider, contextualizes geoengineering within ongoing struggles that civil society is fighting against corporate power and control, exploitation of humans and nature, global inequality, and extractive violence. Fuhr and Schneider argue that geoengineering schemes could shore up corporate and political power for incumbent industries such as fossil fuels, aerospace and defense, agribusiness and mining, and high technology. This suggests a potential convergence of corporate interests in support of the deployment of geoengineering strategies under the current neoliberal regime. Opposed to these powerful interests, the authors also describe civil society's struggles for strong geoengineering governance in various fora and call for a movement of movements against geoengineering technologies.

Next, in chapter 5, Kyle Powys Whyte, a Potawatomi scholar, discusses Indigenous perspectives on geoengineering in a conversation with Buck. He explains how Indigenous persons often are scripted into roles in environmental discussions. This can leave crucial issues unaddressed—Whyte describes how issues like decolonizing treaty rights have been absent from the conversation on Indigenous perspectives on geoengineering. He also explains how frameworks of free, prior, and informed consent (FPIC) are inadequate for participation in geoengineering and advises that instead of trying to bring Indigenous persons into geoengineering discussions, researchers should be thinking about how to make these discussions more meaningful to them. The current situation that geoengineering seeks to preserve, Whyte says, is a dystopia for many Indigenous people or from the perspective of their ancestors, and this makes the starting point not to understand how to preserve this situation but how to get out of it. Fortunately, Indigenous efforts to design governance that approximates ecosystems and reform institutions in ways that can respond to ecosystem dynamics may suggest some paths forward.

In chapter 6, McLaren argues that geoengineering research privileges certain forms of knowledge, expertise, moral theory, and subjectivity that are incompatible with a full account of justice that would recognize all persons (and perhaps other agents) as fully equal moral agents or subjects. By revealing the potential injustices of geoengineering, this contribution steers away from prevailing accounts of geoengineering based in the specific interests of the current Western liberal elites. Instead, it provides a foundation based in recognitional solidarity for obligations to ourselves and our children as well as to entities of the nonhuman world not usually recognized as subjects of justice.

Rounding out this part, Tina Sikka further develops an intersectional analysis of geoengineering in chapter 7. She explains that feminism is constituted by multiple feminisms and discusses two in particular: ecofeminism and standpoint theory. Sikka also describes how structural racism and class could intersect with the development of geoengineering, concluding that "understanding how all these axes of marginalization change and morph with new technologies inclusive of geoengineering is the first step toward reaching a more robust understanding of a whole host of sociotechnical transformations currently underway."

In sum, the second part's emphasis on how geoengineering would perpetuate social inequality and class power dynamics sets the stage for the full-fledged political economy analysis developed in the rest of the book. The third part brings greater emphasis to the capitalist state as the main actor able to address climate change, as its chapters lay out different views on the actual implications of deploying both CDR and SRM strategies from the perspective of state action and economic planning. In chapter 8, Laurence Delina develops a thought experiment—a gedankenexperiment—to imagine how governments around the world would react if they took the immediate threat posed by climate change seriously at this very moment. After arguing that geoengineering entails too high a risk, the chapter explores how the massive mobilization for the transition to renewable energy production that would ensue would likely serve to rationalize and expand state power. In this context of strong state power, how would the dilemma between the mass transition to renewable energy and the deployment of climate geoengineering strategies play out? Delina explores the key implications and trade-offs of this choice as well as the role popular movements would be called to play in these circumstances.

In chapter 9, Christian Parenti builds on the emergency state response context delineated by Delina to hone in on specific interventions progressive governments that recognize the situation could undertake rapidly, mainly the nationalization of CCS and the shutting down of the fossil fuel industry. He explores what an "environmental making state" would look like in actuality so as to delineate a program of action for progressive social movements in the context of the civilizational threat posed by climate change.

In a similar vein, Andreas Malm argues in chapter 10 that the idea of technological manipulation of the climate is premised on an inability to consider state planning of the capitalist economy. However, pushing the analysis further, he argues that minimally reasonable programs of temporary solar geoengineering and/or large-scale deployment of NETs would in themselves demand very extensive planning. Moreover, they can never substitute for the planning required to terminate large-scale fossil fuel combustion. The chapter thus emphasizes the depth of planning required for a successful economic transition to fully address climate change, with or without geoengineering.

In chapter 11, Anne Pasek suggests that we approach geoengineering as infrastructure, which would allow for emphasis to be shifted from production to the reproductive labor of earth systems and their maintenance. Pasek makes a distinction between making and provisioning the climate, pointing out that much critical scholarship falls into the former frame and then discussing the theoretical and political gains from taking up the latter, bringing in theory about care and repair. Such a switch in the framing of geoengineering might better outline the necessity of state planning discussed in previous chapters.

Together, these four chapters outline what a serious state-led attack on climate change would look like in the current crisis context. The fact that this looks nothing like current climate policy strongly suggests that the little action governments have undertaken up to now—establishing uncertain and probably ineffective carbon markets or taxes and embedding reliance on unproven NETs into the Paris Agreement—amounts to little less than a new type of climate change denial.[28]

The fourth and final part examines geoengineering as a class project in its imperialist, class struggle, and political dimensions. It takes on the class dynamics of the capitalist economy so as to understand the deep struggles associated with the development and eventual deployment of geoengineering. Whereas part 3 considered what states *could* do if they took the climate crisis seriously, this last part explores what states might *actually* do, taking into account current developments in the world system and the crisis of neoliberal hegemony. In chapter 12, Richard York scans the *longue durée* of the emergence and development of capitalism and its relation to technological change through the lens of the metabolic rift perspective. He draws a comparison between geoengineering proposals and the now-forgotten project of Atlantropa, designed by German architect Hermann Sörgel during the interwar period. In response to Europe's growing needs for energy and land, Sörgel designed a plan to engineer the whole Mediterranean Basin as well as the river systems down to central Africa through a series of hydroelectric dams, including on the Strait of Gibraltar, that would both generate electricity and free up land by lowering the level of the Mediterranean. The plan was taken seriously by business and political circles, though the eruption of the Second World War prevented its further deployment. This example serves to discuss how technical fixes such as geoengineering and the Atlantropa project have historically been the standard response by capitalists to recurrent crises in the globalizing political economy. However, attempts at fixing problems of the social order through depoliticized technological means that do not address underlying class inequality have typically failed to solve root problems, while at the same time bringing about new ones. More importantly, though, they are also part of an imperialist project of core countries and their capitalist elites to reestablish their power over the world system in times of chaos.

Next, in chapter 13, Kevin Surprise argues from a Gramscian perspective that solar geoengineering should be viewed as a strategy to preempt a coming systemic crisis in which the environmental conditions of the existence of capitalism are being destroyed. Hence as a reaction of capitalism to itself, so to speak, geoengineering is part of the green capitalist passive revolution to ensure the continued hegemony of the ruling classes and capitalist states by maintaining the prevailing political and economic relations currently under threat. Such a framing shifts the debate from consideration of the ethics of geoengineering in terms of affected publics, decision-making, and guidelines for research to one of class struggle in which the dominant elite will hold on to its position and privilege by all means—including deliberately modifying biogeochemical cycles on a planetary scale, likely under the lead of the U.S. military.

In chapter 14, Nils Markusson, Mads Dahl Gjefsen, Jennie Stephens, and David Tyfield, building on their previous work,[29] look at how the "promise" of geoengineering may play out in the current context of growing systemic chaos in core Western countries—as evidenced by the election of Trump, Brexit, and the rise in racism and anti-immigration sentiments in Europe and North America—that might portend the end of neoliberalism. As discussed in other chapters, one of the main concerns is that the mere existence of technological means to address climate change provides a means to accept the reality of the climate crisis all the while shoring up the continued use of fossil fuels. In the scenarios explored in the chapter, a move away from neoliberalism and toward an "illiberal" authoritarianism might open the door to support for solar geoengineering, generally considered as incompatible with neoliberal market commitment. On the other hand, the chapter considers the possibility that China will rise in support of a renewed liberalism as U.S. hegemony declines, with more ambiguous prospects for geoengineering.

Finally, Buck looks in chapter 15 at the new political context of the United States after the election of Trump as president. She looks at the actual messy context in which decisions about geoengineering will likely be made—far removed from the civil rationality of U.N. corridors—which includes, beyond an unpredictable and impulsive head in the most powerful state in the world, the "post-truth" media ecology, meddling in national elections by foreign powers, decreased state legitimacy, and the extreme polarization in national politics as well as in domestic climate discussion. This new context in which climate change denial now looms larger than ever means cuts in funding of climate change research, including geoengineering research, with potentially contradictory consequences: on the one hand, it will undoubtedly delay the development of potentially useful negative emissions technologies, and on the other hand, solar geoengineering deployment could be rendered nearly impossible even though it might be more needed—and perhaps more called for—than ever.

For a Critical Political Ecology of Geoengineering

Overall, the contributions assembled in this book connect the climate crisis with some key themes of our times: race, gender, and class inequality; the election of authoritarian, far-right governments; and historical and current Western and U.S. imperialism. Geoengineering involves much more than simply changing some thermodynamic parameters of an abstract and overheating Earth system in need of management—as a "fever" that needs to be "cured"—as technical discussions imply. The Earth system is made up of myriads of socioecological systems in complex interaction, and all are affected by any modification of global parameters in ways that largely cannot be predicted. The metabolic relationship between humans and the land has been deeply disrupted first by the imperial expansion of European states and the colonization of Indigenous lands and second by the rise of industrial capitalism and steam power—expansionary processes that continue to this day. Artificially created racial divides stemming from imperial expansion transect populations and exacerbate wealth inequality, as do patriarchal relations. Of course, these gender, racial, and colonial lines of fracture are deeply embedded in Western environmental thought.[30] Attempting to understand geoengineering without understanding these socioecological relations and the events that shaped them would leave us with an incomplete picture of the prospect.

The contributors to this volume sought out to show how these lines of fracture and injustice play out in geoengineering discussions so as to start constructing a critical political ecology of geoengineering. They draw the lines between unequal and unjust socioecological relations and geoengineering as a means to extend the current model's life with the effect of further entrenching social inequality. As a case in point, terrestrial CDR under a capitalist system, given its vast land requirements, would likely encroach on lands currently used by smallholders, tenant farmers, and Indigenous people. Who would buy the carbon credits generated by these forests? Polluters in wealthy countries who seek to maintain their line of business, individuals wanting to protect their lifestyles, and states of the Global North seeking to meet their commitments under the Paris Agreement. Solar geoengineering, on the other hand, is specifically designed to preserve existing relations of power by extending the time frame for transitioning capitalist production to renewable electricity, as noted by multiple contributors. As McLaren explains in chapter 6, geoengineering denies the right to be a subject of justice to all but the elite. Sikka and Whyte, in their respective contributions, argue that current geoengineering discussions focused on top-down management of socioecological relations are not relevant to Indigenous people, smallholders, and tenant farmers. We would add further that they do not address the main concerns of the majority of humans for fair access to the land—quite the contrary.

But then, what would a noncolonial response to climate change that included climate engineering look like? Given the scope of the crisis, states would need to play a key role. Hence the third part of the book explores how a state that were to take climate emergency seriously would respond to the current situation. Here perspectives diverge slightly: In chapter 8, Delina argues that states would need to lead a large-scale mobilization of resources to reduce GHG emissions so as to avoid resorting to geoengineering. From a different perspective, Parenti argues in the next chapter that states would be well advised to nationalize CDR technologies so as to take full control of them and ensure rapid deployment. In a related vein, Pasek suggests that viewing geoengineering as infrastructure based on an ethics of care might provide a way to use geoengineering to support life systems instead of bolstering existing unjust structures.

The last theme addressed in this volume is the long-term evolution of the world system in the historical *longue durée*. The neoliberal regime that has dominated the global political economy for the last four decades now appears to be on the brink of implosion,[31] and systemic chaos is at its highest point since the inception of the regime. State governments are nonetheless still deeply committed to free market ideals even though their environmental action (or inaction) will crucially determine whether the crisis is resolved and what regime might subsequently emerge. Overall, this volume outlines the complexity of sociopolitical-economic systems in themselves and of their relation with ecological systems, thereby challenging the widespread accounts of geoengineering deployment that assume the continuity of the liberal order for decades to come.

We do not know what the future will bring. The perspectives in this book, though, demonstrate that geoengineering cannot be reduced to a set of technological options without any political implications or to extrapolations from climate models; it is also not simply a discourse existing only in the minds of researchers, advocates, and policy makers. Geoengineering and climate policy are rather parts of a complex social and ecological system comprising historically situated political and economic structures and processes taking place on a world scale and need to be approached as such.

It is high time to take geoengineering down from the ivory towers and laboratories to the grounds where people live out their conflict-ridden lives. When approaching the brink, we—all of us, not just the experts—need to have some idea of what the options are and what their implementation would mean. This book is our attempt to contribute to the urgently needed conversation and to provide a basis for informed action on the part of social movements, academic researchers, and all citizens. Geoengineering might well be the most momentous technological idea humanity has ever toyed with; it should be a concern for everyone.

Notes

1 Dave Levithan, "Geoengineering Is Inevitable," Earther, October 9, 2018, http://earther.gizmodo.com/geoengineering-is-inevitable-1829623031.

2 Intergovernmental Panel on Climate Change (IPCC), *Global Warming of 1.5°C: An IPCC Special Report on the Impacts of Global Warming of 1.5°C above Pre-industrial Levels and Related Global Greenhouse Gas Emission Pathways, in the Context of Strengthening the Global Response to the Threat of Climate Change, Sustainable Development, and Efforts to Eradicate Poverty* (Geneva, Switzerland: IPCC, 2018).

3 Jonathan Watts, "Geoengineering May Be Used to Combat Global Warming, Experts Say," *Guardian*, October 8, 2018.

4 Cesare Marchetti, "On Geoengineering and the CO_2 Problem," *Climatic Change* 1, no. 1 (1977): 59–68.

5 See James Rodger Fleming, *Fixing the Sky: The Checkered History of Weather and Climate Control* (New York: Columbia University Press, 2011).

6 Mikhail I. Budyko, *Climatic Changes* (Washington, D.C.: American Geophysical Union, 1977).

7 Wallace Broecker, "CO_2 Arithmetic," *Science* 315, no. 5817 (March 2007): 1371; Chris Huntingford and Jason Lowe, "'Overshoot' Scenarios and Climate Change," *Science* 316, no. 5826 (May 2007): 829.

8 The Royal Society, *Geoengineering the Climate: Science, Governance and Uncertainty* (London: Royal Society, September 2009).

9 Royal Society, 49.

10 Royal Society, 49.

11 Royal Society, 49.

12 See, for example, Daniela F. Cusack et al., "An Interdisciplinary Assessment of Climate Engineering Strategies," *Frontiers in Ecology and the Environment* 12, no. 5 (2014): 280–287.

13 Silke Beck and Martin Mahony, "The Politics of Anticipation: The IPCC and the Negative Emissions Technologies Experience," *Global Sustainability* 1 (2018): e8; Roger Pielke, Tom Wigley, and Christopher Green, "Dangerous Assumptions," *Nature* 452, no. 7187 (2008): 531–532; Wim Carton, this volume, chap. 3.

14 For example, solar geoengineering does not reduce ocean acidification. See H. Damon Matthews, Cao Long, and Ken Caldeira, "Sensitivity of Ocean Acidification to Geoengineered Climate Stabilization," *Geophysical Research Letters* 36, no. 10 (2009).

15 David Keith, *A Case for Climate Engineering* (Cambridge, Mass.: MIT Press, 2013).

16 Jane A. Flegal and Aarti Gupta, "Evoking Equity as a Rationale for Solar Geoengineering Research? Scrutinizing Emerging Expert Visions of Equity," *International Environmental Agreements: Politics, Law and Economics* 18, no. 1 (2018): 45–61.

17 Duncan McLaren, "Framing Out Justice: The Post-politics of Climate Engineering Discourses," in *Climate Justice and Geoengineering: Ethics and Policy in the Atmospheric Anthropocene*, ed. Christopher J. Preston (Lanham, Md.: Rowman & Littlefield, 2016), 139–160.

18 Mike Hulme, *Can Science Fix Climate Change? A Case against Climate Engineering* (Cambridge, U.K.: Polity, 2014); see also Stephen Gardiner, "Geoengineering and Moral Schizophrenia: What Is the Question?," in *Climate Change Geoengineering: Philosophical Perspectives, Legal Issues, and Governance Frameworks*, ed. Wil C. G. Burns and Andrew L. Strauss (Cambridge: Cambridge University Press, 2013), 11–38.

19 James Rodger Fleming, "Climate Engineering and Surgeons," *Environmental History* 19, no. 2 (April 1, 2014): 281–364; see also Clive Hamilton, *Earthmasters: The Dawn of the Age of Climate Engineering* (New Haven, Conn.: Yale University Press, 2013).

20 Timothy W. Luke, "Geoengineering as Global Climate Change Policy," *Critical Policy Studies* 4, no. 2 (2010): 111–126.

21 Holly Jean Buck, "Climate Engineering: Spectacle, Tragedy or Solution? A Content Analysis of News Media Framing," in *Interpretive Approaches to Global Climate Governance: Deconstructing the Greenhouse*, ed. Chris Methmann, Delf Rothe, and Benjamin Stephan (New York: Routledge, 2013), 166–180.

22 See, for example, Corinna Dengler and Lisa Marie Seebacher, "What about the Global South? Towards a Feminist Decolonial Degrowth Approach," *Ecological Economics* 157 (2019): 246–252; Giorgos Kallis, *Degrowth* (Newcastle upon Tyne: Agenda, 2018); Ståle Holgersen and Rikard Warlenius, "Destroy What Destroys the Planet: Steering Creative Destruction in the Dual Crisis," *Capital & Class* 40, no. 3 (2016): 511–532; Ashish Kothari, Federico Demaria, and Alberto Acosta, "Buen Vivir, Degrowth and Ecological Swaraj: Alternatives to Sustainable Development and the Green Economy," *Development* 57, nos. 3–4 (2014): 362–375; and Christine Bauhardt, "Solutions to the Crisis? The Green New Deal, Degrowth, and the Solidarity Economy: Alternatives to the Capitalist Growth Economy from an Ecofeminist Economics Perspective," *Ecological Economics* 102 (2014): 60–68.

23 See, for example, Joshua B. Horton and David W. Keith, "Solar Geoengineering and Obligations to the Global Poor," in *Climate Justice and Geoengineering*, 90.

24 Royal Society, *Geoengineering the Climate*; see also Kevin Anderson and Glen Peters, "The Trouble with Negative Emissions," *Science* 354, no. 6309 (October 14, 2016): 182–183.

25 Stephen M. Gardiner, "Is 'Arming the Future' with Geoengineering Really the Lesser Evil? Some Doubts about the Ethics of Intentionally Manipulating the Climate System," in *Climate Ethics: Essential Readings*, ed. Stephen M. Gardiner Simon Caney, Dale Jamieson, and Henry Shue (Oxford: Oxford University Press, 2010), 284–312; Christopher J. Preston, "Ethics and Geoengineering: Reviewing the Moral Issues Raised by Solar Radiation Management and Carbon Dioxide Removal," *Wiley Interdisciplinary Reviews: Climate Change* 4, no. 1 (2013): 23–37.

26 Hamilton, *Earthmasters*.

27 Augustin Fragnière and Stephen M. Gardiner, "Why Geoengineering Is Not 'Plan B,'" in *Climate Justice and Geoengineering*.

28 William Carroll et al., "The Corporate Elite and the Architecture of Climate Change Denial: A Network Analysis of Carbon Capital's Reach into Civil Society," *Canadian Review of Sociology* 55, no. 3 (2018): 425–450.

29 Nils Markusson et al., "The Political Economy of Technical Fixes: The (Mis)Alignment of Clean Fossil and Political Regimes," *Energy Research & Social Science* 23 (2017): 1–10.

30 Pascale d'Erm, *Soeurs en écologie: Des femmes, de la nature et du réenchantement du monde* (Nantes, France: La mer salée, 2017); Kyle Powys Whyte, "Way beyond the Lifeboat: An Indigenous Allegory of Climate Justice," in *Climate Futures: Reimagining Global Climate Justice*, ed. Kum-Kum Bhavnani et al. (Oakland: University of California Press, 2019); Ariel Salleh, *Ecofeminism as Politics: Nature, Marx and the Postmodern* (London: Zed Books, 2017).

31 See, for example, Alain Lipietz, "Fears and Hopes: The Crisis of the Liberal-Productivist Model and Its Green Alternative," *Capital & Class* 37, no. 1 (2013): 127–141.

Part II

Contesting Geoengineering

Power, Justice, and Civil Society

2

Winning Hearts
and Minds?

●

Explaining the Rise of the
Geoengineering Idea

INA MÖLLER

Over the past few years, geoengineering has become an idea to be reckoned
with. Although floating around as an obscure example of scientific hubris for
more than forty years, it has only been a decade or so since the idea has come
to settle on the minds and desks of scientists, philosophers, and policy makers. But within these past ten years, the topic has become so potent that media
celebrities like Fareed Zakaria are broadcasting the subject into my parents'
living room and that governments around the world feel the need to form a
political position. Funnily enough, most of the suggestions out there are arguably still closer to science fiction than they are to reality. The question that
I address in this chapter is how geoengineering was able to rise to the global
agenda.

Is This the End?

Most often, geoengineering's uncanny appeal is attributed to a growing sense
of environmental emergency and the daunting amount of mitigation needed
to avoid a climate catastrophe. Its increasing presence within the climate
change conversation is understood as the direct response to evidence about
already occurring climate change impacts (more droughts, more floods, more
hurricanes), the continuous resistance of states and industries to reducing
greenhouse gas (GHG) emissions, and the recognition that even extreme
mitigation would not be enough to avoid severe impacts of climate change.[1]
Others have pointed to the widespread disillusionment with international
climate politics and the increase of voices, including those of respected scientists, that call for serious consideration of geoengineering technologies.[2]
Undoubtedly, these things play an important role in promoting the concept
and enabling its diffusion. Without rising temperatures and repeated failures

of politicians to provide a convincing solution, the idea to actively intervene in the earth's atmosphere would have no motivation, no audience, and no resonance. There would be no reason to speak about geoengineering if we weren't aware of the physical and political problems that shape the landscape of climate change policy.

Nevertheless, it is misleading to think of geoengineering as a natural or inevitable response to the current crisis. Scientific evidence of climate change and widespread skepticism toward climate politics constitute a necessary precondition for geoengineering to be discussed, but alone, they are not sufficient to explain why it has become so potent. After all, the geoengineering idea is competing with a wide array of alternative suggestions to deal with climate change. Some of these are decidedly neoliberal, matching the paradigms of economic growth and the primacy of the market in the form of carbon trading schemes. Others come out of a more transformative background, including the movement to divest from fossil fuels. Still others have merged ideas from different backgrounds and given life to popular concepts like the circular economy. What these suggestions have in common with geoengineering is that they evolved in the same time frame, are part of the same context of international politics, and have access to the same scientific information. Geoengineering is not the only inevitable outcome of the contemporary climate crisis. As I will argue in this chapter, it is the result of a special set of circumstances connected to the way we study climate change, the particular authority of (physical) science in this field, and the creation of a well-integrated knowledge community.

Global Solutions for Global Problems

Geoengineering as a whole is usually suggested and advanced by scientists with a very particular view on the climate—namely, a planetary one. As the English saying goes, "If you have a hammer, every problem looks like a nail." In this case, one could say that if you have a global climate model, the solutions you think of are likely to look like some form of geoengineering. Both negative emissions and solar geoengineering stem from work that relies on global climate models. These models use parameters like the earth's reflectivity and its capacity to absorb CO_2 in order to forecast what our own increase of GHG emissions might do to global temperatures. Naturally, if the parameters can be altered on the computer screen, it is not so far a stretch to suggest altering them in real life. Meanwhile, the costs and consequences of this alteration for societies or ecosystems are often too complex or unpredictable to include in the calculation, which is why geoengineering technologies often turn out to be the most "cost-effective" options available. (Wim Carton explains the role of models and their assumptions in further detail in chapter 3.)

The latest result of this sort of calculation can be seen in the Intergovernmental Panel on Climate Change's (IPCC's) special report *Global Warming of 1.5 °C*, which spells out a need to invest in large-scale negative emissions if we want to have any hope of staying below 2°C or 1.5°C average warming. The models used to calculate these results are integrated assessment models (IAMs), called this because they integrate an economic cost-benefit model with a model of the earth's physical systems. Their purpose is to come up with the cheapest possible option by which we can stay below a given temperature target. But IAMs generally do not include any of the messy political details and local circumstances that differentiate one area from another. To the model, desert is desert, rainforest is rainforest, and field is field—no matter whether these areas are in Bangladesh, Afghanistan, or the United States. They also assume that future generations will be richer than contemporary generations, meaning that future generations should be able to pay more for climate policies than contemporary generations. The blindness of the models toward local circumstance and intergenerational equity almost automatically results in the large-scale deployment of technologies that are indiscriminately placed around the world. Their inbuilt assumptions lead to the inescapable conclusion that we will need large amounts of negative emissions technologies (NETs) to avoid the dangerous effects of climate change. Similarly, the view of the planet as a system that can be modeled and adjusted on a screen makes solar geoengineering seem like a comparatively feasible and straightforward option. Global climate models and the way we "see" climate change through them have played an important role in defining what the problem is and consequently provided geoengineering as a solution. But while global models created the mindset for the creation of geoengineering, there are a couple of further mechanisms that helped bring geoengineering onto a political agenda.

Making Geoengineering Governable

In 1992, the U.S. National Academy of Sciences (NAS) published a report that could be regarded as one of the first steps in the process of making geoengineering into a governable object. Although individual geoengineering techniques had already been discussed in a small number of publications during the 1970s and '80s,[3] the attention that the 1992 report brought to geoengineering technologies initiated what James Fleming calls the third (and contemporary) "cycle of promise and hope" for technologies to engineer the climate.[4] In its analysis of the policy implications of climate change, the 1992 NAS report essentially created what we are now talking about by devoting a section to "geoengineering," defining it as "options that would involve large-scale engineering of our environment in order to combat or counteract the effects of changes in atmospheric chemistry."[5]

The inclusion of geoengineering in this report was not undisputed. Stephen Schneider, one of the NAS report's panel members, remembers that "the very idea of including a chapter on geoengineering led to serious internal and external debates. Many participants (including myself) were worried that even the very thought that we could offset some aspects of inadvertent climate modification by deliberate climate modification schemes could be used as an excuse by those who would be negatively affected by controls on the human appetite to continue polluting and using the atmosphere as a free sewer."[6]

In the end, however, the skeptics were convinced by Robert Frosch, lead author of the section and at that time vice president of the General Motors

Research Lab in Michigan, that the dramatic reduction of fossil fuel use or the endurance of catastrophes due to climate change constitute such horrors that geoengineering might be the "least evil" option. The fear of a looming apocalypse prevailed over all other concerns, enabling the topic to be included in the report and (to this date) providing the main argument against not engaging in geoengineering research.

Frosch's analysis included a list of large-scale environmental engineering options, including massive afforestation, sunlight reflection methods, enhanced ocean carbon uptake, and direct capture of CO_2 from the air. Among these, stratospheric aerosol injection, the increase of cloud abundance, and ocean iron fertilization (OIF) were considered most promising based on estimated cost and effectiveness. Social, legal, or ethical consequences were not discussed, and the need for research was focused primarily on the risks and side effects that the interventions might have in terms of natural systems. The report constituted the first resemblance of what we understand as geoengineering today and set the scene for scientific investigation.

The NAS's appraisal of geoengineering technologies was followed by a series of scientific investigations that revolved, in particular, around OIF. The year 1993 saw the first experiment (IronEx I), in which scientists from Moss Landing Marine Laboratories in the United States fertilized a 64 km^2 patch of ocean near the Galápagos Islands to create an algal bloom. This experiment was soon followed by others, including research institutes from New Zealand, the United Kingdom, Canada, Japan, and Germany, who conducted experiments in the Southern Ocean and the North Pacific Ocean. Meanwhile, in a special issue of the journal *Climatic Change*, physicists and atmospheric chemists interested in altering the earth's reflectivity argued that climate models could be used to test the approach based on the recommendations that the NAS report had given.[7] Their assessments were accompanied by publications that took a more societal perspective, also picking up the NAS's geoengineering agenda and bringing it into the realm of ethics and international law.[8] Overall, however, investigation into geoengineering options continued with a relatively low profile, and much of the debate was restricted to corridors of scientific conferences.

Around 2006/2007, several things happened that took the geoengineering idea to another level. On the OIF front, Russ George, a private entrepreneur who had founded the company Planktos Corp., announced that he would use OIF at a commercial scale to generate carbon credits. Aiming to dissolve 100 tons of iron over an area of 10,000 km^2, the company planned to conduct the first of a series of experiments on the high seas off the Galápagos Islands. This plan caused an outcry by the Ecuadorian government and brought OIF to the international agenda. In 2007, the United States, Greenpeace, and the International Union for Conservation of Nature (IUCN) submitted papers to a meeting of the parties to the London Convention and London Protocol on ocean dumping. Their "statement of concern" resulted in a legal process (further whipped up by the actual deployment of OIF by George in 2012) that led to an amendment of the London Protocol in 2013. The amendment prohibited ocean fertilization for nonscientific purposes, and the incident essentially hoisted the topic to an unprecedented level of international concern. It also led to multiple decisions within the U.N. Convention on Biological Diversity (CBD) several years later.[9]

The year 2006 also saw the publication of another special issue in the journal *Climatic Change*—this time including an editorial essay by Nobel Prize–winning scientist (and coiner of the Anthropocene concept) Paul Crutzen. Essentially reiterating what the NAS report and earlier publications had suggested, Crutzen argued that the distribution of sulfate particles in the earth's atmosphere could provide a solution to the policy dilemma of climate change. In a back-of-the-envelope calculation, he suggested that spraying 1–2 million metric tons of sulfur per year into the stratosphere at a cost of $25–$50 billion[10] would be enough to offset the effects of global warming. This, he argued, would provide an alternative to the difficulties of coordinated mitigation and the stalemate of international negotiations.[11] Crutzen's high-level scientific profile brought many scientists who were previously concerned about being associated with the topic out of the closet and opened the floodgates for publications on geoengineering.[12]

Incensed on the one hand and encouraged on the other, critics and supporters of Crutzen's argument took the essay as a cue to engage more vocally with the geoengineering idea, in particular with his suggestion to use sulfur injections. In parallel to more modeling exercises, much of the debate revolved around questions concerning what technologies constituted geoengineering, why (or why not) geoengineering was worth researching, and if research should look into this type of technology at all. Some of the titles published during this time give a sense of the types of discussions going on: "20 Reasons Why Geoengineering May Be a Bad Idea," "Will Desperate Climates Call for Desperate Geoengineering Measures?," and "Geoengineering: What, How, and for Whom?"[13]

The many question marks associated with the concept eventually led the U.K. Royal Society to publish a comprehensive report on the science, governance, and uncertainty of geoengineering. Endorsed by one of the oldest and most prestigious scientific authorities in the world, twelve scientists managed to bring an unprecedented level of order to the chaotic geoengineering discussion, setting the terms, priorities, and definitions that would shape further research endeavors and publications.[14] Its expressed aim was to "provide an authoritative and balanced assessment of the main geoengineering options" in a subject area "bedevilled by doubt and confusion."[15] The report was influential in the sense that it provided the motivation and legitimation for several coordinated research programs on geoengineering and was followed by governmental inquiries in the United Kingdom, the United States, and Germany.

Apart from further legitimizing the importance of geoengineering research with its name, the report's most significant act was to categorize geoengineering technologies according to their underlying physical mechanisms. Geoengineering was divided into carbon dioxide removal (CDR) and solar radiation management (SRM), with CDR labeled as relatively unproblematic in terms of side effects and social considerations and SRM labeled as highly risky and in need of additional regulatory mechanisms.

Notably, the distinction between physical mechanisms was not the only obvious demarcation available. Steve Rayner, one of the two social scientists on the committee, remembers that there had been a dispute about how to group differ-ent geoengineering technologies. Next to the physical categorization, there were other suggestions, including a distinction between black-box engineering (con-tained technologies like direct air capture [DAC]) and technologies that would intervene in open ecosystems. Although the CDR and SRM categories made a lot of sense from a climate system perspective, questions like scale, jurisdiction, or areas of intervention would have been more aligned with the common prin-ciples of environmental law. These suggestions were, however, overruled by the predominantly natural science makeup of the scientists involved, and CDR and SRM remain the main categorizations for inquiries into the legal and institu-tional aspects of geoengineering today.

The physical process-based demarcation resulted in two effects. The first was that it created a bridge for geoengineering to become a part of the mainstream climate science discussion. Thanks to the CDR and SRM categories, the main defining aspect of geoengineering—scale—became obscured by the microlevel mechanisms of carbon sequestration and reflectivity potential that were being studied. Differences between the original geoengineering suggestions (ocean fertilization, stratospheric aerosol injection) and newer, more "benign" sugges-tions (using biochar to enhance carbon uptake in soil, growing light-colored crops) became blurred and subsumed under the same umbrella term. The result of these developments was that geoengineering's mainframe was no longer a list of hubristic emergency plans for planetary temperature management. At least among those scientists who began to seriously engage with the issue, it became a respectable area of inquiry within earth systems science.

The second effect was the introduction of governance to the geoengineering debate. The critical discussions about social dimensions of geoengineering were answered by a recognition that proper management would be needed, thereby assuming that geoengineering was indeed a manageable or governable object. In the wake of the report, much of the social science literature on geoengineering (and SRM in particular) began engaging with governance, thereby moving the focus from a question of essence ("Should" or "could" we engineer the plane-tary climate?) to a question of process and design (How do we manage research and/or deployment?). What exactly this governance should look like remains unclear, but suggestions range from designing a scientific code of conduct to the creation of an international governmental oversight body.

"Invisible Colleges" and the Network Effect

The increase of voices calling for research on geoengineering is often used to explain why geoengineering is rising on the agenda. But who are these voices, and did they all independently conclude that geoengineering might be neces-sary? Social network analysis, interviews with scientists, and observations at geoengineering research conferences reveal some of the social dynamics that influence the discussions around it and that added to its diffusion.

An effective way of getting people involved in a discussion that they have never been part of in the sphere of science is by inviting them to conferences, preferably with a title that is sexy enough to raise people's curiosity and with enough funding to cover their travel costs. In 2010, the Climate Response Fund—an organization dedicated to the support of research on geoengineering and financed through the industry-backed philanthropy Guttman Initiatives—had invited scientists, engineers, and lawyers to think about governance mechanisms that would ensure safe and responsible research into climate engineering techniques. The conference was held at the Asilomar Center in Monterey Bay, California, to mirror a famous earlier conference that had created a set of voluntary guidelines to deal with the hazardous research of recombinant DNA. By choosing this setting, the scientific organizers (most of whom had already been involved in geoengineering research for some years) hoped that the conference would end in a similar set of guidelines to answer the governance needs spelled out by the Royal Society's report. Coincidentally, having a code of conduct would also be useful in avoiding the imposition of external regulation.

This did not work out quite as well as expected. Many invited researchers were not at all familiar with geoengineering yet, nor did they share a necessary level of agreement. After several days of intense discussions on whether to subsume CDR techniques under the same umbrella as spraying particles into the atmosphere and what was considered acceptable outdoor experimentation, the organizing committee decided to recommend the adoption of an already existing set of principles (the so-called Oxford Principles[16] suggested by an ad hoc working group linked to the Royal Society report), although no formal endorsement of these principles took place.[17]

All in all, the conference outcome was perhaps not as clear as originally desired, but it did have the effect of opening the research field to more people. For many attendants, Asilomar was the first time they engaged with geoengineering at all, and some of these would continue working in the field for years to come. One of the main conclusions made in the conference report was that geoengineering would need continuous discussion and involvement of new voices to ensure that research and governance development would be inclusive. This reasoning has continued to motivate the organization of international conferences and workshops until today. Since Asilomar, more than seventy-five workshops, conferences, and summer schools dedicated exclusively to geoengineering have taken place around the world, reaching a couple of thousand attendees. With this kind of diversity and activity, it seems indeed that there are many voices involved. But what does this really mean for the discussions themselves?

Interesting insights can be gained by looking at the public data provided by geoengineering events (organizers, speakers, attendance lists). An analysis of the conference conveners shows that two-thirds of the conferences and workshops are part of a network of organizations repeatedly involved in setting up geoengineering events, with some especially active think tanks and advocacy groups—the Institute for Advanced Sustainability Studies (IASS) in Germany, the Center for International Governance Innovation (CIGI) in Canada, the Solar Radiation Management Governance Initiative (SRMGI) in the United Kingdom, and the Environmental Defense Fund (EDF) in the United States—acting as links for bringing together other initiators and funders. This casts a more nuanced light on the assumption that geoengineering is becoming a generally pressing issue, showing that the organization of platforms through which the topic is presented and discussed in the first place is driven by a relatively small set of actors.

It also becomes clear that despite a considerable number of participants overall (more than 2,500 attendees participating in seventy-one events over eleven years),[18] the number of recurring individuals—meaning those who show their commitment to the topic by repeatedly attending and speaking at these meetings—is just a fraction of this. Only about 20 percent attended more than one event, narrowing down the community of actively engaged individuals to about 330 people. Within this network, there are again considerable differences in dedication, with two-thirds constituting what I would call an "interested periphery" that might work on geoengineering to a limited degree and a "dedicated core" of about one hundred people who are much more actively involved. Most of the organizations that these people are affiliated with are based in the Global North, primarily in the United States, the United Kingdom, and Germany.[19]

Such numbers are not unusual. Scientists who study scientific networks also call them "invisible colleges": informal communication networks of approximately one hundred elite scholars from different affiliations who confer power and prestige to one another through citations, coauthorships, and common projects.[20] While there are different ways of mapping them, the main characteristics that define invisible colleges are that they work on the same issue (clustered around different subcategories), that they find means and ways to meet regularly, and that they collaborate to gain funding. The existence of an invisible college around geoengineering is further supported by impressions from scientists who have been part of the network for many years and who are familiar with most of the people working in this area. The effect of being a part of the network is that people get to know and like each other. They form a common identity. Direct criticism of ideas or research approaches becomes more difficult to express, and getting to know each other on a personal level dampens initial skepticism toward the geoengineering idea. One of the researchers I interviewed described this development in the following way:

> I remember when I went to Asilomar for the first time as a young professional and I was expecting to be put in this place with Machiavellian scientists who want to manipulate the Earth's atmosphere. I was scared of them before I'd even met them, and of course it's very disarming when you get to know people. I have to try to keep that kind of distance, but I think especially the atmospheric scientists are just terrified of political inaction. I feel that they're personally motivated by the right reasons, and therefore surely their research could be good. But I'm also aware that it can have fairly broad-reaching implications. I guess in a way it's putting faith in future regulation. But if we look at how well or not well we've done regulating climate change and emissions, we really shouldn't believe in that.... Rationally, I should be terrified.

One interesting thing about this reflection is that geoengineering technologies were originally suggested as a way to circumvent the lethargy and ineffectiveness of climate change governance, yet they themselves are now expected to require similar, if not higher, levels of political engagement and regulation.

Although many network members are aware of the possible dangers associated with geoengineering and with stratospheric aerosol injection in particular, their proximity to the research and to the people involved seems to obscure some of the reservations that they might otherwise feel. Coming closer to the research object makes it feel safer than when looking at it from far away. This is probably another reason much of the discussion that takes place in the core network is focused on technical issues (modeling approaches, governance mechanisms) rather than fundamental questions of social feasibility or desirability.

Basic assumptions about what geoengineering is, why it is important, and how it should be researched also serve as important functions for community building. This becomes particularly clear in situations where "new" actors with an alternative understanding of the issue area join the discussion. At one of the larger conferences dedicated to the subject, a small group of nongovernmental organization (NGO) representatives had organized a session that discussed "radical alternatives" (i.e., nongeoengineering) pathways to climate change policy. They argued that all alternative solutions to climate change would need to be ticked off before geoengineering warranted serious discussion and that the sheer idea of governing something like stratospheric aerosol injection was absurd, as it was already difficult enough to govern climate change itself (Lili Fuhr and Linda Schneider, in chapter 4, and Kyle Whyte, in chapter 5, discuss similar arguments).

Prominent scientists from the core geoengineering network countered the NGO presentation heavily with allegations of having ideological and unrealistic ideas of alternative pathways. They argued that ecological farming, renewable energy, and behavioral change would never be enough to save the carbon emissions required to maintain a healthy climate. While the NGO representatives talked about oppression, capitalism, and the creation of military-industrial complexes, the scientists talked about inadequate carbon sequestration, rising populations, and responsibility toward the public for not exploring all possible solutions. Tempers rose and the event escalated into scientists and NGO representatives accusing each other of not answering questions and not listening to answers.

In the aftermath of this incident, I had a short conversation with another social scientist who falls into my "dedicated core" category of the network analysis and who turned out to have a completely different opinion of what we had both just witnessed. While I thought the scientists had been rather hostile and intentionally leading the conversation to questions of technical detail (e.g., "How much carbon can a small-scale farmer sequester on a square meter of land?"), my conversation partner perceived the NGO representatives as obnoxious, naïve, and uninformed: "It's always the same with them." I have also heard these types of comments on other occasions, indicating that there is a very clear perception of who is "us" and who is "them."

In this context of clashing worldviews, a productive discussion can become difficult. Yet both sides are influential in their own ways, and as one side advocates for more engagement with geoengineering, the other advocates for less. The result is confusion among policy makers and the public. Is it good? Is it necessary? Do we need it? Should we avoid it? Differences in underlying values and problem definitions lead to different conclusions, and the conflicts between nonstate actors are likely to find repetition among state actors when these start engaging with geoengineering in a policy making context.

Introducing Geoengineering to Climate Politics

How does all this relate to the arrival of geoengineering on the political agenda? The power of science to influence global policy is hotly discussed in both international relations and science and technology studies. Debates revolve mainly around the direction and degree of influence, with the frontier being between those who see science and politics as separate spheres and those who see the two as coconstituting each other.[21] But whatever position one chooses to take, there can be little doubt that the way science defines problems and solutions has an important role to play in the shaping of world politics. Particularly in an issue area as technical and far removed from everyday experience as climatic change, policy is dependent on scientific information to provide the basis for political negotiation.

For scientific ideas and results to become relevant for international politics, they need to be backed by a wide community and preferably endorsed by an authoritative body like the IPCC. In most nations, this organization is seen as the primary authoritative actor in the field of climate science, and policy makers refer to its reports for information about issues linked to climate change. Yet the IPCC does not do "research" in the conventional sense. Its role is to assess the existing climate literature and summarize it in a way that is useful to a nonscientific audience.

The first key to having any scientific issue introduced to international climate discussions is therefore publishing enough material to make it into the IPCC's integrated assessment reports, something that the geoengineering research community has been very good at. After the increase of geoengineering publications following Crutzen's essay and the Royal Society report, the 2014 IPCC report on climate change included the first substantial discussion of geoengineering (in the form of CDR and SRM) as a potential policy option, although it still expressed great caution toward reliance on SRM techniques. Following the line of the Royal Society report and subsequent publications, CDR was treated as largely benign and synonymized with the idea of "negative emissions"—a concept that had evolved in parallel literature around the idea of bioenergy with carbon capture and storage (BECCS). (See Carton's arguments in chapter 3 for more detail.)

As a consequence of the merger, those geoengineering scholars who work primarily with CDR have become increasingly affiliated with the negative emissions community, and solar geoengineering is turning into a separate group. Where this development will lead is still unclear, but it is not necessarily the most helpful distinction for policy making. With CDR and negative emissions subsuming all approaches that remove carbon from the atmosphere (from reforestation and restoration of natural sinks to the large-scale, industrial use of BECCS), it becomes difficult to distinguish between sustainable policy approaches that support ecosystems and unsustainable approaches that may lead to the further destruction of ecosystems. Similarly, because SRM subsumes all types of enhanced reflectivity, planting brighter crops or increasing local cloud cover is always associated with the more problematic characteristics of stratospheric aerosol injection. Although these distinctions may seem like technicalities, the definition of geoengineering is crucial to international law and politics, and political negotiations are likely to revolve around clarifying these distinctions before they can go on to regulating them.[22]

The second key for science to make it into political discussions is political fit, meaning that a scientific issue is framed in such a way that it matches already ongoing discussions and can be subsumed under existing frameworks and concepts. In climate change negotiations, the most prominent concepts shaping the agenda are mitigation and adaptation, whereas mitigation (the reduction of GHG emissions) is more popular among developed nations, and adaptation (the preparation for the effects of unavoidable climate change) is more popular among developing nations. Increasingly, the CDR-type geoengineering techniques are labeled as mitigation measures, thereby improving their political acceptability.[23] This kind of redefinition is also visible in the IPCC's special report *Global Warming of 1.5°C*, where CDR and negative emissions are only once defined as a form of geoengineering and referred to as a form of mitigation throughout the rest of the report. Even among SRM researchers, emphasis is starting to be laid on the potential for SRM to reduce carbon concentrations in the atmosphere.[24] At the same time, SRM is sometimes described as a form of climate adaptation, and several prominent scientists argue that it will be particularly beneficial to developing countries and those nations who are most vulnerable to climate change.[25] This type of framing is fundamentally different from earlier discussions, where geoengineering was traded as a third response different from both mitigation and adaptation and allegedly independent of protracted political negotiations.

Political fit is being created not only in the context of climate policies but also in the area of security. Particularly solar geoengineering (mostly discussed in the form of stratospheric aerosol injection) is often presented as a serious threat to global security if deployed by a unilateral actor, therefore requiring a global regulatory institution. At the same time, the technology is also discussed as the only viable option in case of a climate emergency, therefore requiring research and development. This powerful narrative of heroes and villains who could save or destroy the world speaks to a wide audience and might convince some to take up the question of geoengineering governance more earnestly.

How this governance should take place remains unclear and is already subject to conflict. In March 2019, the Swiss government—backed by a diverse coalition of states from around the world—introduced a draft resolution on the governance of geoengineering to the U.N. Environment Assembly (UNEA). This resolution suggested that geoengineering technologies, both CDR and SRM, should be examined more closely by the secretariat of UNEA in order to provide a better knowledge basis for decision-making. But because the resolution also suggested that the global community take a precautionary approach toward research and development, several powerful oil- and gas-producing countries—including the United States, Brazil, and Saudi Arabia—blocked the resolution in its early stages. Although negotiations are likely to continue, this is the first indicator that the conflicting worldviews observed among nonstate actors are likely to play a prominent role as political discussions on this topic evolve.

In less than two decades, scientific investment and authority transformed geoengineering from fringe science to a politically discussed option in the catalog of climate change policies. It was initially brought up under the assumption that transitioning away from a fossil fuel–based economy would be less feasible than engineering the earth's atmosphere, and it became acceptable thanks to the repeated endorsement of respectable scientific actors. But far from being an automatic response to climate change, it was subject to a lengthy debate within scientific research that, until a few years ago, was mostly about whether to discuss it at all. What has not changed in that time are core concerns that have plagued the idea from the very beginning: fears of creating a slippery slope from research to deployment, the possibility that geoengineering research could distract from an urgent reduction of GHG emissions, the desirability and usefulness of outdoor experiments, and the viability of using computer models to predict effects and side effects. All these topics shaped the conversation at Asilomar in 2010, and they are still prominent in the conversation today. The answer to why this is the case lies in the nature of the issues. They are questions of value rather than questions of knowledge. Because values will continue to play a major role as the geoengineering idea continues to spread, the decision-making process is unlikely to be any easier or more straightforward than conventional mitigation or adaptation policies.

Notes

1 Edward A. Parson, "Climate Engineering in Global Climate Governance," *Transnational Environmental Law* 3, no. 1 (2014): 89–110; Martin L. Weitzman, "A Voting Architecture for the Governance of Free-Driver Externalities, with Application to Geoengineering," *Scandinavian Journal of Economics* 117, no. 4 (2015): 1049–1068.

2 Wil Burns and Simon Nicholson, "Governing Climate Engineering," in *New Earth Politics: Essays from the Anthropocene*, ed. Simon Nicholson and Sikina Jinnah (Cambridge, Mass.: MIT Press, 2016), 343–366.

3 Cesare Marchetti, "On Geoengineering and the CO_2 Problem," *Climatic Change* 1, no. 1 (1977): 59–68; Hubert H. Lamb, "Climate-Engineering Schemes to Meet a Climatic Emergency," *Earth Science Reviews* 7, no. 2 (1971): 87–95; Freeman J. Dyson and Gregg Marland, "Technical Fixes for the Climatic Effects of CO_2," in *Carbon Dioxide Effects Research and Assessment Program: Workshop on the Global Effects of Carbon Dioxide from Fossil Fuels*, ed. William P. Elliott and Lester Machta (Springfield: National Technical Information Service [NTIS], Department of Commerce, 1979), 111–118.

4 James Rodger Fleming, "The Pathological History of Weather and Climate Modification: Three Cycles of Promise and Hope," *Historical Studies in the Physical and Biological Sciences* 37, no. 1 (2006): 3–25.

5 National Academy of Sciences (NAS), *Policy Implications of Greenhouse Warming: Mitigation, Adaptation, and the Science Base* (Washington, D.C.: National Academy Press, 1992).

6 Stephen H. Schneider, "Geoengineering: Could—or Should—We Do It?," *Climatic Change* 33 (1996): 295.

7 Robert E. Dickinson, "Climate Engineering: A Review of Aerosol Approaches to Changing the Global Energy Balance," *Climatic Change* 33 (1996): 279–290.

8 Daniel Bodansky, "May We Engineer the Climate?," *Climatic Change* 33 (1996): 309–321; Dale Jamieson, "Ethics and Intentional Climate Change," *Climatic Change* 33 (1996): 323–336.

9 For a more detailed analysis of the entrepreneurial activities of Russ George and their effects on his local partners, see Kate Elizabeth Gannon and Mike Hulme, "Geoengineering at the 'Edge of the World': Exploring Perceptions of Ocean Fertilisation through the Haida Salmon Restoration Corporation," *Geo: Geography and Environment* 5, no. 1 (2018): e00054.

10 The symbol $ refers to U.S. dollars.

11 Paul J. Crutzen, "Albedo Enhancement by Stratospheric Sulfur Injections," *Climatic Change* 77 (2006): 211–219.

12 Paul Oldham et al., "Mapping the Landscape of Climate Engineering," *Philosophical Transactions of the Royal Society A* 372, no. 2031 (2014): 20140065.

13 Robert A. Frosch, "Geoengineering: What, How, and for Whom?," *Physics Today* 62, no. 2 (2009): 10; Barbara Goss Levi, "Will Desperate Climates Call for Desperate Geoengineering Measures?," *Physics Today* 61, no. 8 (2008): 26; Alan Robock, "20 Reasons Why Geoengineering May Be a Bad Idea," *Bulletin of the Atomic Scientists*, May–June 2008, 14–59.

14 Aarti Gupta and Ina Möller, "De Facto Governance: How Authoritative Assessments Construct Climate Engineering as an Object of Governance," *Environmental Politics* 28, no. 3 (2019): 480–501.

15 The Royal Society, *Geoengineering the Climate: Science, Governance and Uncertainty* (London: Royal Society, 2009).

16 Steve Rayner et al., "The Oxford Principles," *Climatic Change*, 121 (2013): 499–512.

17 Jeff Goodell, "A Hard Look at the Perils and Potential of Geoengineering," Yale Environment 360, April 1, 2010, http://e360.yale.edu/content/feature.msp?id=2260.

18 The total number of attendees is probably significantly higher, as some events might not have been captured in the database and not all events provide full lists of their participants. However, the overall conclusions regarding actively engaged participants are not affected by this.

19 Frank Biermann and Ina Möller, "Rich Man's Solution? Climate Engineering Discourses and the Marginalization of the Global South," *International Environmental Agreements* 19, no. 2 (April 2019): 151–167.

20 Alesia Zuccala, "Modeling the Invisible College," *Journal of the American Society for Information Science and Technology* 57, no. 2 (November 2005): 152–168.

21 Rolf Lidskog and Göran Sundqvist, "When Does Science Matter? International Relations Meets Science and Technology Studies," *Global Environmental Politics* 15, no. 1 (2015): 1–20.

22 For example, the Convention on Biological Diversity (CBD) has made a nonbinding but authoritative decision to prohibit any deployment of geoengineering that affects biodiversity, with the exemption of small-scale scientific studies. Depending on what falls under the definition of *geoengineering*, this decision could have serious impacts for different technologies and approaches.

23 Olivier Boucher et al., "Rethinking Climate Engineering Categorization in the Context of Climate Change Mitigation and Adaptation," *WIREs Climate Change* 5, no. 23 (2014): 23–35.

24 David W. Keith, Gernot Wagner, and Claire L. Zabel, "Solar Geoengineering Reduces Atmospheric Carbon Burden," *Nature Climate Change* 7 (2017): 617–619.

25 Jane A. Flegal and Aarti Gupta, "Evoking Equity as a Rationale for Solar Geoengineering Research? Scrutinizing Emerging Expert Visions of Equity," *International Environmental Agreements: Politics, Law and Economics* 18, no. 1 (2018): 45–61.

3

Carbon Unicorns
and Fossil Futures

———————————————————●

Whose Emission Reduction
Pathways Is the IPCC
Performing?

WIM CARTON

If one is to believe recent Intergovernmental Panel on Climate Change (IPCC) reports, then gone are the days when the world could resolve the climate crisis merely by reducing emissions. Avoiding global warming in excess of 2°C/1.5°C now also involves a rather more interventionist enterprise: to remove vast amounts of carbon dioxide from the atmosphere, amounts that only increase the longer emissions refuse to fall.[1] The basic problem with this idea is that the technologies that are supposed to deliver these "negative emissions" do not currently exist at any meaningful scale. Given the large uncertainties surrounding their feasibility; their expected effects on land use change, food security, and biodiversity; and their scalability, it moreover seems improbable that they ever will.[2] Indeed, there appears to be something of an unspoken consensus among scientists that the mitigation scenarios represented in the IPCC increasingly mirror science fiction writing. In a recent assessment, the European Academies Science Advisory Council (EASAC), for example, concluded that negative emissions technologies (NETs) have "limited realistic potential" to help mitigate climate change on the scale that many scenarios assume will be needed.[3] One expert summarized the skepticism well when she recently characterized such technologies as "carbon unicorns,"[4] underscoring the widening gap between the level of mitigation that is needed and the apparent infeasibility of the pathways that are supposed to take us there.

Despite its fantastical nature, however, the negative emissions idea has recently burst into the public arena, where it is already leading a life of its own. For skeptics, this raises the concern of a "moral hazard," or the possibility that the mere promise of future NETs could act as a break on emission reductions in the present.[5] Techno-optimist policy makers, the thinking goes, might very well seize on the negative emissions idea as a "get-out-of-jail" card, holding back from rapid near-term decarbonization in the belief that opportunities for future negative emissions offer a sufficient guarantee that the climate crisis can be contained. It is, above all, future generations, and particularly the poorest among them, that would face the consequences when this "high-stakes gamble" eventually backfires and large-scale NETs turn out to be little more than a pipe dream.[6] At that point, the window of opportunity for avoiding dangerous warming through conventional mitigation would have closed, and the world would be left with the unenviable choice between runaway warming or implementing some of the more dystopian geoengineering technologies that this book documents. These are not empty fears: as I discuss next, the perceived necessity to defer the bulk of mitigation into a discounted future is the exact logic that underpins the rise to prominence of NETs in mitigation scenarios.[7] How can we expect policy makers to guard against wishful thinking when even scientists appear unable to do so? Besides, the negative emissions concept has already strayed beyond the realm of abstract science and policy debates. The business case for mitigation deferral is already under construction, suggesting that NETs are already performing valuable political-economic work. This makes it necessary to scrutinize much more closely what is actually going on in the various models that generate the apparent need for negative emissions.

Take the example of Shell. While not exactly known for its vanguard mitigation actions, the company recently released a document in which it outlines its vision to keep global warming "well below 2°C."[8] Unsurprisingly perhaps, Shell's "most ambitious climate scenario" turns out to include substantial fossil fuel use well into the future. It, for example, assumes that demand for oil will grow until about 2025 and then decrease only gradually. By 2050, the year when the world needs to reach net-zero emissions in order to stay below 1.5°C,[9] oil demand in this scenario would still account for about 85 percent of current consumption. By 2070, the net-zero target for 2°C, fossil fuel production would still be responsible for 16.5 $GtCO_2$, or almost half of what it is today. For Shell to be able to claim that these estimates are compatible with the targets of the Paris Agreement, it heavily relies on speculative technologies, in particular carbon capture, usage, and storage (CCUS) and NETs. It thus assumes that all the remaining fossil fuel carbon can be captured and/or compensated for by storing it in products (6.1 $GtCO_2$/yr), applying direct carbon capture and storage (CCS) to oil and gas installations (3.4 $GtCO_2$/yr), and deploying large-scale bioenergy with carbon capture and storage (BECCS—6.1 $GtCO_2$/yr), which is the NET most often favored in models. In total, this would require that "some 10,000 large carbon capture and storage facilities are built, compared to fewer than 50 in operation in 2020."[10] To reach 1.5°C, the company then imagines that an additional effort could be made by planting "another Brazil in terms of rainforest."[11]

These astonishing claims fulfill a clear function, even if they are only a scenario exercise or a best-case "possible" future, not a concrete prediction or commitment. The inclusion of NETs and CCUS in Shell's future scenario constructs a vision in which the risk for stranded assets is minimized. It makes it possible to claim, as Shell does in its *Shell Energy Transition Report*, that all the company's proven and potential fossil fuel reserves could be utilized—around twenty-five years of reserves at current production rates—while still staying within the limits of the Paris Agreement.[12] Invoking a future of large-scale negative emissions in this way suggests that there is no need to cut fossil fuel production before its economic value has been fully recovered and thus no need for drastic short-term changes in the company's business model.[13] Given the urgency of the climate problem, this surely seems extraordinary. Is Shell making these numbers up? An analysis by Carbon Brief suggests that the math does indeed add up. Despite being somewhat optimistic about future energy demand in general, Shell's projections of future coal, oil, and gas demand and of the scale at which NETs could be deployed are all broadly in line with those of 2°C-compatible IPCC scenarios. If anything, Shell's scenario is at the lower end of how much negative emissions models say could be deployed by the end of the century.[14]

In itself, of course, it is unremarkable that a fossil fuel company would use all means possible to help justify the continued use of oil and gas, including fostering narratives about the large-scale deployment of future carbon unicorns. This, after all, is the company that has known about the dangers of climate change since at least the 1980s and still decided to double down on oil and gas investments.[15] More surprising is the fact that this logic appears fully internalized in mainstream climate scenarios—in other words, that the IPCC reports appear to feature emission reduction pathways that seem fully compatible with massive continued fossil fuel use in the medium term. More than a "moral hazard," this suggests some fairly hazardous scientific morals. Surely this should raise a few eyebrows. How is it possible that the world's most authoritative science on climate change is generative of scenarios that play directly in the hands of the fossil fuel industry? In this chapter, I want to explore some of the reasons this is occurring. I want to argue that the path that led to the inclusion of negative emissions in models and from there into the IPCC was a profoundly ideological one and that we need to understand it as such to make sense of the way in which negative emissions are already being invoked to justify business as usual. Doing so, I suggest, helps us challenge the now common idea that negative emissions are somehow an inevitable reality of climate politics.

Negative Emissions as Convenient Fiction

To unpack the work that negative emission scenarios perform, we need to start with the science that produces them. The scenarios represented in the IPCC are generated by using so-called integrated assessment models (IAMs), which are designed to model the complex relationship between social and biophysical systems.[16] Briefly put, these models seek to project future technological innovation, economic growth, demographic change, energy use, and so on and how these interact with changes in the climate system. A first important observation is that economics plays a central role in this exercise, in that IAMs are generally made to operate in line with mainstream economic theories. The IPCC is quite explicit about what this means. The fifth assessment report (AR5), for example, notes that "the models use economics as the basis for decision making. This may be implemented in a variety of ways, but it fundamentally implies that the models tend toward the goal of minimizing aggregate economic costs of achieving mitigation outcomes. . . . In this sense, the scenarios tend towards normative, economics-focused descriptions of the future."[17] The IPCC also acknowledges that models "typically assume fully functioning markets and competitive market behavior" and therefore do not take account of existing asymmetries and (market) power relations.[18]

This focus on economics is important for a number of reasons. Most directly, it means that climate policy in IAMs is interpreted as the implementation of a carbon price—that is, it is the assumed cost of carbon that gives the main incentive for a specific level of mitigation. Other mechanisms by which transformational change might come about—for example, through mass behavioral changes or nonmarket government interventions on the scale of recent Green New Deal proposals—are largely ignored by the models.[19] A second and related constraint lies in the cost-minimization focus that the IPCC mentions. Essentially, IAMs are designed to "maximize overall welfare" and find the most cost-effective emission reduction pathways. This effectively means that they prioritize different mitigation technologies on the basis of primarily economic and technological criteria and underplay social, political, and broader environmental reasons society might opt for one mitigation technology over another.[20] In fact, this is the main reason a technology like BECCS can be modeled by IAMs on such obviously unrealistic scales (e.g., requiring a land area twice the size of India). Even when models take into account more explicitly social factors (e.g., to assess the public acceptability of different technologies), these are usually still translated into economic terms.[21]

Now, this primary concern in IAMs with optimized, cost-effective mitigation pathways has long meant that very few scenarios were compatible with keeping temperatures below 2°C. Up to the fourth assessment report or so, models tended to generate results that stabilized greenhouse gas (GHG) concentrations at levels that were significantly higher than those corresponding with what are now the Paris Agreement targets.[22] As political recognition of the need for a 2°C limit grew, first in Europe and then elsewhere, policy makers asked the modeling community to come up with scenarios that would be consistent with this.[23] This confronted modelers with a considerable dilemma. As Parson notes, "Most of the Integrated Assessment Models (IAMs) . . . found that the target could not be met via plausible and cost-effective levels of mitigation."[24] The solution they came up with was as innovative as it was problematic. Modelers decided to include in IAMs novel mitigation options, primarily BECCS and afforestation, that allow for the removal of CO_2 from the atmosphere. These were not entirely conjured out of thin air, of course. Afforestation had long been promoted as a carbon offsetting strategy, and researchers had put forward the possibility for BECCS already in the late 1990s and early 2000s, though it had so far only been considered as a "backstop" option. Now, however, it became the go-to method.[25] Not only did this significantly decrease the costs of achieving stringent mitigation targets;[26] it also introduced a debt mechanism into the models.[27] By allowing for large-scale carbon dioxide removal (CDR), it suddenly became possible to exceed carbon budgets in the short term, on the assumption that this "overspending" would be compensated for by net-negative emissions in the second half of the twenty-first century.[28]

The inclusion of NETs in IAMs in this way played a crucial role in upholding the possibility of the 2°C limit. As Dooley et al. argue, "The availability of

BECCS proved critical to the cost-efficiency, and indeed the theoretical possibility, of these deep mitigation scenarios, leading to systemic inclusion of BECCS in RCP2.6 scenarios [the IPCC's most optimistic scenarios] included in AR5."[29] It is worth underscoring what this means. NETs were mainstreamed in IAMs in order to square the request of policy makers (i.e., to provide 2°C pathways) with the specific economic framework within which these models operate. Current scenarios are in this sense the result of a cost-minimization exercise[30]—a fully institutionalized effort to keep the costs of mitigation as low as possible. The models are therefore not actually telling us that NETs are a biophysical necessity to achieve stringent mitigation targets. They are merely saying that these technologies are *more cost-effective than other forms of mitigation*. Whether one accepts the need for negative emissions in this sense ultimately depends on whether one agrees with the various economic assumptions upon which the models are based. As I discuss next, there are plenty of reasons not to do so.

The Politics of a Pathway

Modelers tend to see their work as "objective input[s] to the climate policy debate,"[31] as do, presumably, most policy makers. They are generally quite candid about the assumptions that underpin their models but insist that scenarios are still useful because they are not actually meant to be policy prescriptive or offer accurate predictions of the future. Rather, modelers argue, scenarios are merely supposed to be policy relevant to "support policy decisions between different choices" and point to those pathways that would be most efficient.[32] The IPCC has in many ways sought to patrol this border between policy-relevant and policy-prescriptive science.[33]

A rich literature in science and technology studies, however, suggests that this distinction is difficult to uphold in practice. Scholars in this discipline point out that any kind of scientific knowledge production comes with value judgments and therefore inevitably ends up fulfilling some kind of political function.[34] The incorporation of NETs in IPCC scenarios is one clear illustration of how, as Turnhout et al. put it, "dominant political discourses compel scientists to create assessments that work within these discourses,"[35] a process that involves the articulation of problems that are legible to and the proposal of solutions compatible with prevailing political and economic logics. Knowledge production, in other words, is often reflective of existing power relations in society, and at the same time, it contributes to and justifies the reproduction of those relations. The future focus and therefore unverifiable and speculative character of scenario production significantly amplifies these dynamics.[36] In this, the problem is not that science is political per se but that its political character remains unrecognized or actively denied by the actors involved, either directly or as a consequence of the methods that are used. As a result, value-laden and contestable assumptions appear as somehow unavoidable or "natural," which closes opportunities for debate and the involvement of dissenting voices. The use of models, particularly ones as complex as IAMs, further contributes to this process of depoliticization by shrouding assumptions and value judgments behind seemingly technocratic and objective modeling choices.[37]

Beck and Mahony argue that the increasing importance of modeled emission reduction pathways in the IPCC in this way represents a shift toward a "new politics of anticipation, wherein potentially contestable choices for climate futures are woven into the technical elaboration of alternative pathways."[38] They note that by being included in the authoritative assessments of the IPCC, such pathways do not just describe possible climate futures but potentially help bring them into being—that is, they *perform* certain futures as seemingly legitimate, necessary, and desirable. IAMs in this sense provide scientific backing for the kind of mitigation scenarios that are "thinkable and therefore actionable"[39] while simultaneously sidelining others. One of the clearest examples of this is the negative emissions idea. Before they appeared in IAMs, NETs were virtually absent from the climate policy arena. Following their inclusion in models, they appeared in IPCC assessments and from there have become an increasingly common topic in mainstream policy debates. As the previous Shell example shows, they have now moved into the delaying tactics of the fossil fuel industry. The modeling community in this way "performed an important legitimating function for the speculative technology of BECCS, pulling it into the political world, making previously unthinkable notions . . . more mainstream and acceptable, as well as perhaps pushing it ahead of policy options (such as radical mitigation) in political calculations."[40] The speculative and contestable inclusion of NETs in influential and seemingly neutral IPCC assessments served to normalize and mainstream the idea that negative emissions are both feasible and necessary.

Taking this one step further, some scholars have argued that the negative emissions idea is performing an important legitimizing role for the existing architecture of climate policy as a whole.[41] By perpetuating the idea that cost-effective pathways to 2°C and now also 1.5°C are still available, the argument goes, the IPCC is providing a rather convenient narrative to governments. The possibility of future NETs appears to suggest that more of the same incremental policies will eventually get us there—that there is no need for drastic or economically "irrational" actions.[42] As such, it helps preserve a sense of normality against increasingly dire warnings—and observations—of an unfolding climate emergency against thirty years of political delay in delivering serious mitigation efforts. The science-sanctioned normalization of negative emissions in this sense reproduces the idea that all is as it should be in the magical wonderland of climate politics, where mitigation need not imply efforts to cut actual fossil fuel production, at least not in the short term. At the same time, this discourse builds on highly improbable projections of the future that rely on the hypothetical deployment of technologies that—at the scale they are being proposed—reasonably belong in the realm of science fiction. When it so obviously constitutes a form of risk transfer, in which it is the powers that be that stand to gain, while it is future generations that will be left to pick up what pieces remain,[43] then the need for critique runs very deep indeed.

Performing the Imperative of Gradualism

So how did it come to this? To understand how IPCC scenarios ended up being "performative" in this way requires us to scrutinize not just model outcomes and the political work that these perform but also the logics that generate these outcomes in the first place. There is plenty to suggest that the dynamics described in the science and technology literature can in large part be traced back to the various connected assumptions that underlie IAMs, assumptions that together constitute an ideological commitment to the postulates of mainstream economic theory. This is, of course, hardly a unique case. In important ways, it reflects the wider trend by which economics has come to dominate the terms of the climate policy debate—of how to assess and understand both the problem and its potential solutions.

Consider again the focus of IAMs on cost-effective mitigation. Why exactly is it that the prioritization of cost-effective solutions leads to the need for negative emissions? There are a number of intertwined reasons for this, and while I cannot consider all of them here, a few stand out as particularly important. First, it is worth noting that mitigation costs in IAMs are usually calculated on the basis of a comparison with a so-called baseline, meaning a counterfactual scenario of what the world would look like in the absence of climate policies. The cost of mitigation, in other words, is an estimate of what it takes, in economic terms, to move from the assumed baseline to the desired mitigation scenario. Observe that these baselines are necessarily hypothetical exercises, not in the least because, with a few exceptions, models so far do not take into consideration the many feedbacks of a warming climate itself.[44] Essentially, they assume that economic growth, population growth, consumption, energy demand, and so on will continue as an extrapolation of existing trends despite rapidly increasing temperatures, as if climate change has no societal impact at all. This crucial omission is acknowledged by modelers as a shortcoming, but in itself, it arguably already invalidates the entire scenario-building exercise. Calculating costs and cost-dependent mitigation pathways in relation to an impossible baseline clearly overstates the benefits of the "no-policy" scenario and therefore presumably inflates the aggregate costs of mitigation. More generally, it means that the choice of baseline significantly influences the outcomes of the model.[45] Modelers generally deal with this by considering a large range of possible baselines that are grouped together under stylized "socioeconomic pathways."[46]

To different extents, these baseline scenarios assume continued (and often growing) fossil fuel consumption and trade well into the twenty-first century.[47] Moving to a mitigation scenario, then, logically implies significantly reducing that consumption and trade as well as its corresponding economic value (since baselines are seen as economically optimal, any deviation from them becomes a cost). The extent to which fossil fuel consumption needs to be reduced, however, and the exact costs this corresponds to fundamentally depend on the kind of mitigation technologies that are included in the model. For example, if one assumes a future in which no CCS technologies are implemented, then fossil fuel consumption needs to fall rapidly to stay within the targeted temperature limits, reaching zero before the end of the century.[48] Indeed, many of the scenarios that explicitly exclude CCS (including BECCS) are unable to generate 2°C-compatible pathways at all because of prohibitively high costs.[49] This reflects not only the substantial investments needed to rapidly replace current high-carbon infrastructures but also the fact that for many sectors where there are currently few low-carbon technological alternatives on the horizon—think cement and steel production, aviation, and so on—drastic emission cuts would almost, by necessity, involve cuts in economic production. With CCS, some of those fossil fuels can continue to be used and their corresponding economic value recovered. The inclusion of negative emissions from BECCS in particular extends this effect further. BECCS essentially enlarges the carbon budget while also providing a source of energy, allowing even more fossil fuels to be used in the medium term.[50] Observe here that the cost-effective focus of IAMs in this way renders different mitigation technologies qualitatively substitutable, meaning that as long as a given technology is available and economically attractive (within the assumptions used by the model), it will be prioritized. As noted previously, this ignores obvious social justice or environmental sustainability concerns.

From this discussion, it appears that the cost of mitigation tends to decrease with the continued use of more fossil fuels. This is obviously not fully true. As the IPCC points out, aggregate mitigation costs in IAMs generally *increase* when action is delayed.[51] The reason for this is fairly simple—scenarios still need to reach 2°C or 1.5°C by the end of the century. The longer mitigation is delayed, the more fossil fuels are "locked into" a (growing) economy and the more investments and/or devaluations it will therefore take to eventually bring emissions down to net zero / net negative. The cost of mitigation is hence a function not of continued fossil fuel use per se but of the steepness of the mitigation curve—that is, of how quickly fossil fuel consumption needs to fall in order to reach the specified temperature target. The faster fossil fuels are eliminated, the steeper the emission reduction curve and therefore the higher the cost. This seems like a trivial consideration, but it is critical to understand its implications. Since IAMs are designed to minimize mitigation costs, this means that they *by definition* select for the most gradual reduction in fossil fuel use. As long as emissions and fossil fuel consumption go hand in hand, this also means that they select for the most gradual emission reduction curve. Including CCS in IAMs essentially decouples fossil fuel consumption from emissions and therefore allows the former to fall more slowly relative to the latter. Negative emissions go even further in that they actually extend the carbon budget and thus stretch out the emission reduction curve itself. The effect is to reduce the rate at which fossil fuel use needs to fall, which in turn leads to lower mitigation costs. One could say that the inclusion of NETs in IAMs in this way serves to recover as much economic value from fossil fuel consumption and trade as possible within the limits of a 2°C or 1.5°C budget.

Some of this "gradualizing" of the mitigation curve is done quite explicitly by modelers themselves. Van Vuuren et al.,[52] for example, using an earlier version of the IAM IMAGE, explain the criteria they used when developing their mitigation pathways: "First, a maximum reduction rate was assumed reflecting the technical (and political) inertia that limits emission reductions. Fast reduction rates would require the early replacement of fossil fuel–based capital stock, and this may involve high costs. Secondly the reduction rates compared to baseline were spread out over time as far as possible—but avoiding rapid early reduction rates and, thirdly, the reduction rates were only allowed to change slowly over time."[53]

Kriegler et al.,[54] using a different IAM, similarly note that their model does not allow for the early retirement of existing fossil fuel infrastructure. In other words, the models are actively designed so as to avoid the devaluation of economically valuable fossil fuel assets, believing this to be unfeasible and so as to make full use of the window of opportunity for reaching the desired mitigation target. In this, their assumptions are directly in line with the arguments of the fossil fuel industry. In Shell's "well below 2°C" scenario as well, the imperative for NETs logically follows from the assumed inevitability of socioeconomic and technological inertia—that is, the idea that until 2030 or so, "energy system CO_2 emissions are largely locked in by existing technologies, capital stock, and societal resistance to change."[55] Modelers and industry interests in this way agree that there is no alternative to incremental change, even if that means conjuring up improbable technological solutions.

These dynamics are reinforced by the idea that future costs and benefits need to be discounted relative to the present. IAMs generally use a discount rate of 5 percent,[56] which means they weigh costs and benefits in the present more heavily than those that will occur in the future. The reasoning here, imported directly from financial markets, is that future generations will be wealthier (given continued economic growth) than current generations and will therefore be better able to pay for any future costs that arise from climate change. This is a contentious and oft-debated assumption. For one, it assumes, wrongly, that the costs and benefits of mitigation/adaptation and indeed the impacts of climate change itself can be straightforwardly captured / compensated for in monetary terms. As discussed earlier, it also suggests that growth can and will continue despite an accelerating environmental crisis, which seems improbable to say the least. There is furthermore no consensus among economists about what exact discount rate to use, which is unsurprising given the inherently subjective and speculative nature of the exercise.[57] As Stanton et al. note, selecting a discount rate essentially means making a judgment about how to value the benefits of avoided warming for future generations, which is "a problem of ethics, not economic theory or scientific fact."[58] A high discount rate is an implicit prioritization of short-term interests over long-term ones, or as Jasanoff pointedly says, it "erases the distant future as a topic of calculable concern."[59] In the IAMs we are concerned with here, applying a discount rate of 5 percent has the effect of deferring mitigation costs into the future when those costs will supposedly be more affordable. Because large-scale NETs are projected to be implemented mainly in the second half of the century, discounting makes them comparatively more attractive than mitigation measures that are rolled out in the near term and therefore gives them a direct advantage in the model.

So what is actually going on here? Clearly, the supposed necessity of negative emissions in mitigation scenarios is the result of a number of specific assumptions and value judgments, all of which can reasonably be questioned. But the problem seems broader than just the negative emissions issue alone. Essentially, what is being performed in IPCC scenarios is the *imperative of gradualism*—that is, the idea that mitigation needs to be incremental if it is to materialize at all. The "naturalization" of fossil fuel benefits through business as usual baselines, the management of the *rate of mitigation* by way of cost-effective technology choices, the direct "gradualization" of model inputs, and the application of a high discount rate all form the idea that some degree of emissions is inevitable and that the economic benefits of fossil fuel production must be defended to the greatest extent possible. Models in this way institutionalize the assumption that short-term devaluation of fossil fuel assets is untenable and economically undesirable and hence that the socioeconomic inertia is an unavoidable feature of the current energy system. This de facto enacts inertia as some kind of natural law rather than a condition that is maintained and reproduced through historically specific socioeconomic structures and therefore responsive to political choice.

Connecting integrated assessment modeling to the interests of polluters like Shell, then, is a commitment to the ideology of mainstream economics, a narrow reliance on cost-effectiveness as the most appropriate way to mediate between alternative climate futures. By reducing mitigation to a question of carbon costs and then applying a cost-minimization model to it, IAMs render climate change mitigation legible to vested political and economic interests but at the same time delimit the range of mitigation options that seem feasible. As a result, modeled pathways end up being biased against more radical, near-term emission reductions, opportunities for widespread behavioral changes, or the kind of state-driven economic planning proposed by Andreas Malm in this book.[60] It then becomes more logical to imagine that warming will be contained by a massive rollout of fantastical NETs than to try to project, for example, a portfolio of more short-term and risk-averse strategies, even if that means accepting a higher economic cost (for some!). By placing IAM-based scenarios center stage in its assessments, the IPCC in this way reproduces the idea that it is the (contestable and flawed) laws of economic theory that should determine the rules of engagement in climate policy, not the laws of the biogeochemical carbon cycle or consideration for the ethical distribution of mitigation risks and responsibilities. The inevitable end result, ironically, is that the IPCC, as the most authoritative international body on climate change, is providing scientific backing for the kind of delaying tactics that companies like Shell excel in.

The Point Is to Change It

To be sure, there are plenty of good reasons to support certain kinds of CDR, at least in principle. Afforestation is direly needed not just to sequester carbon but also to bend the trend of rapid biodiversity loss. Soil carbon sequestration not only takes carbon out of the atmosphere but also increases soil organic matter and therefore improves soil structure, helps build soil fertility, and benefits soil organisms.[61] Neither of these, however, are the silver bullets that IPCC scenarios are projecting with NETs. Implementing these technologies on a planetary scale comes with enormous challenges, and it therefore seems problematic to treat them as real alternatives to direct emission cuts. In fact, no new research is needed to demonstrate that afforestation, bioenergy production, and CCS are not the convenient and inexpensive mitigation options that they are now being portrayed as. These technologies already exist at smaller scales and have already been extensively studied. The vast literature on carbon forestry, for example, not only confirms the potential benefits that tree planting offers but also vividly illustrates the trade-offs commonly involved, including a real possibility for violence and dispossession, project failure, public disapproval, or the marginalization of the interests and voices of those most affected.[62] Debates on forest-based carbon offsetting—a mechanism that in many ways overlaps with the logic of negative emissions—furthermore underscore the ethical problems with the idea that land use change should compensate for the continued emissions of fossil fuels. Fairhead et al. in this context speak of the "economy of repair," or the idea that "unsustainable use 'here' can be repaired by sustainable practices 'there,'"[63] where "there" often ends up meaning the developing world, since the "economy of repair" too is a cost-optimizing one. If large-scale negative emissions provide the next frontier for this perverse logic, as seems a real risk, it needs to be challenged and resisted.

I have suggested that a good place to start this task is by scrutinizing the idea that negative emissions are necessary in the first place. It turns out that NETs were introduced in models first and foremost as an economic necessity, given by the character of the models themselves. Whether we accept the inevitability of negative emissions—at scale—is therefore entirely contingent on whether we subscribe to the economic assumptions that they extend from. These assumptions ultimately revolve around the treatment of climate change as primarily a question of cost-minimizing economics. It seems obvious that this is a wholly inadequate way to decide on the most feasible, desirable, or appropriate way to cut emissions. It falsely constructs all forms of mitigation as qualitatively equal (ignoring important ethical, political, and ecological differences[64]), perpetuates simplistic assumptions of how change occurs in complex social systems, and orients the mitigation curve toward gradualism despite the social and environmental risks this entails. The cost of mitigation in models is moreover a constructed category fully dependent on assumed long-term technology costs, the exclusion of climate feedbacks, and the choice of discount rates and baselines. Translating this inherently partial approach into concrete mitigation pathways seems like high-risk theoretical myopia and ends up ignoring real opportunities for more just and immediate cuts in GHG emissions. Modelers might insist that their scenarios are not predictions, but their inclusion in the IPCC still gives them undue real-world validity and political influence. It is illuminating in this respect that Van Vuuren et al. recently published a study that modeled scenarios to 1.5°C with minimal negative emissions simply by assuming more rapid electrification of the energy system and far-reaching lifestyle changes, among other things.[65] While they don't provide a cost analysis for these scenarios, one can assume that they would be significantly more costly—in the way IAMs assess this—than "standard" mitigation approaches. What this illustrates is that if one tinkers long enough with inputs and assumptions, it is possible to make these models come up with virtually anything. As Tavoni and Socolow note, this "should make the reader cautious about carrying modeling results into the real world."[66]

In the end then, while modelers acknowledge that the choice between different mitigation options remains a political one, their models only give credibility to a select range of options. By reducing climate policy to a question of cost-optimization, IAMs appear to take the cost of mitigation outside of the political debate. They seem to suggest that mitigation needs to be cost-effective if it will materialize at all, which underplays both the scope and the urgency of the change that is needed. The need for rapid, radical emission reductions suggests a need to repoliticize discussions on what forms of mitigation are most appropriate and how we will be paying for it. Surely, if the responsibility of the IPCC's working group on mitigation extends beyond minimizing the devaluation of fossil fuel assets—as of course it does—then its work should involve highlighting, in a much more direct way, the benefits of certain emissions reduction pathways *in spite* of their cost—that is, to illuminate the many uncertainties and risks of incremental climate policy. Surely assessing opportunities for mitigation should involve not just acquiescing to the inevitability of fossil-infused inertia but actively challenging it by providing an open and honest evaluation of the social, economic, political, and environmental pros and cons of the full range of mitigation options, including those that are inconvenient to vested political and economic interests.

Of course, some economists would fume that no such thing is possible; that high-cost scenarios are politically unrealistic, not policy-relevant; and that no politician or business would implement a policy that is not cost-effective. But that would be missing the point entirely. As Alyssa Battistoni rightly observed recently, there are no politically realistic climate change mitigation options.[67] There is nothing politically realistic about assuming that large-scale NETs are going to save the day. It merely defers the political inconvenience of implementing those technologies to future generations, pushing the problem out of sight for the current generation of decision makers. To accept this as a matter of fact is to fail to stand up to the magnitude of the challenge, to default on our collective responsibility toward future generations. It is to deny that the only realistic way forward involves a fundamental change of politics. Moreover, even *if* it were true that political decisions are necessarily made in narrowly defined, cost-optimizing ways and hence that the political arena is locked into long-term socioeconomic inertia, why should scientists have to play by that game? Why would modelers need to build political feasibility into their models if all this does is lead to future scenarios populated by carbon unicorns? Why should the academic community not point out that there is in fact a choice here, even if it is an unpopular and economically difficult one? When climate policies turn out to be so woefully inadequate, it is perhaps time for the scientific community to become a little less policy relevant and a little more confrontational in its engagement with decision makers.[68] It is perhaps time to start refusing to perform, through seemingly innocuous models, the kind of gradualism that has long ago proven incapable of taking us out of this mess.

Notes

1   Carl-Friedrich Schleussner et al., "Science and Policy Characteristics of the Paris Agreement Temperature Goal," *Nature Climate Change* 6 (July 2016): 827–835; Glen P. Peters and Oliver Geden, "Catalysing a Political Shift from Low to Negative Carbon," *Nature Climate Change* 7 (2017): 619–621; Intergovernmental Panel on Climate Change (IPCC), *Global Warming of 1.5°C: An IPCC Special Report on the Impacts of Global Warming of*

1.5°C above Pre-industrial Levels and Related Global Greenhouse Gas Emission Pathways, in the Context of Strengthening the Global Response to the Threat of Climate Change, Sustainable Development, and Efforts to Eradicate Poverty (Geneva, Switzerland: IPCC, 2018); IPCC, *Climate Change 2014: Mitigation of Climate Change; Working Group III Contribution to the Fifth Assessment Report of the Intergovernmental Panel on Climate Change* (Geneva, Switzerland: IPCC, 2014).

2 See Kevin Anderson and Glen Peters, "The Trouble with Negative Emissions," *Science* 354, no. 6309 (2016): 182–183; Pete Smith et al., "Biophysical and Economic Limits to Negative CO_2 Emissions," *Nature Climate Change* 6 (2016): 42–50; Alice Larkin et al., "What If Negative Emission Technologies Fail at Scale? Implications of the Paris Agreement for Big Emitting Nations," *Climate Policy* 18, no. 6 (2017): 690–714; Sabine Fuss et al., "Betting on Negative Emissions," *Nature Climate Change* 4, no. 10 (2014): 850–853; Anna B. Harper et al., "Land-Use Emissions Play a Critical Role in Land-Based Mitigation for Paris Climate Targets," *Nature Communications* 9, no. 1 (2018).

3 European Academies Science Advisory Council (EASAC), *Negative Emission Technologies: What Role in Meeting Paris Agreement Targets? EASAC Policy Report* (Salle, Germany: EASAC, 2018).

4 Matt McGrath, "Caution Urged over Use of 'Carbon Unicorns' to Limit Warming," BBC News, October 5, 2018, http://www.bbc.com/news/science-environment-45742191.

5 Nils Markusson, Duncan McLaren, and David Tyfield, "Towards a Cultural Political Economy of Mitigation Deterrence by Negative Emissions Technologies (NETs)," *Global Sustainability* 1 (2018): e10; Dominic Lenzi, "The Ethics of Negative Emissions," *Global Sustainability* 1 (2018): e7; Jan C. Minx et al., "Negative Emissions: Part 1—Research Landscape, Ethics and Synthesis," *Environmental Research Letters* 13, no. 6 (2018): 063001.

6 Henry Shue, "Climate Dreaming: Negative Emissions, Risk Transfer, and Irreversibility," *Journal of Human Rights and the Environment* 8, no. 2 (2017): 203–216; Anderson and Peters, "Trouble with Negative Emissions."

7 Minx et al., "Negative Emissions."

8 Royal Dutch Shell, PLC, *Shell Scenarios: Sky—Meeting the Goals of the Paris Agreement* (The Hague, Netherlands: Royal Dutch Shell, PLC, 2018).

9 IPCC, *Global Warming of 1.5°C.*

10 Shell, *Shell Scenarios*, 6.

11 Adam Vaughan, "Shell Boss Says Mass Reforestation Needed to Limit Temperature Rises to 1.5 C," *Guardian*, October 9, 2018, http://www.theguardian.com/business/2018/oct/09/shell-ben-van-beurden-mass-reforestation-un-climate-change-target.

12 Shell, *Shell Energy Transition Report*, 2018, https://www.shell.com/energy-and-innovation/the-energy-future/shell-energy-transition-report.html.

13 Wim Carton, "'Fixing' Climate Change by Mortgaging the Future: Negative Emissions, Spatiotemporal Fixes, and the Political Economy of Delay," *Antipode* 51, no. 3 (2019): 750–769.

14 Simon Evans, "In-Depth: Is Shell's New Climate Scenario as 'Radical' as It Says?," Carbon Brief, March 29, 2018, http://www.carbonbrief.org/in-depth-is-shells-new-climate-scenario-as-radical-as-it-says.

15 Damian Carrington and Jelmer Mommers, "'Shell Knew': Oil Giant's 1991 Film Warned of Climate Change Danger," *Guardian*, February 28, 2017, http://www.theguardian.com/environment/2017/feb/28/shell-knew-oil-giants-1991-film-warned-climate-change-danger.

16 Note that there is also a different set of IAMs that is used to calculate the social cost of carbon and is not used in producing emission reduction pathways. These more simple models make a cost-benefit analysis of different emission reduction pathways by weighing the economic costs of various mitigation options against the risks (again, in economic terms) of climate change. This is the kind of thinking that, for example, leads William Nordhaus—using his Dynamic Integrated model of Climate and the Economy (DICE model)—to the conclusion that the economically "optimal" level of warming is

somewhere from 2.6°C to 3.5°C and that "the advantage of geoengineering over other policies is enormous." See William D. Nordhaus, "An Optimal Transition Path for Controlling Greenhouse Gases," *Science* 258, no. 5086 (1992): 1315–1319; William D. Nordhaus, *A Question of Balance: Weighing the Options on Global Warming Policies* (New Haven, Conn.: Yale University Press, 2008); William D. Nordhaus, "Projections and Uncertainties about Climate Change in an Era of Minimal Climate Policies," *American Economic Journal: Economic Policy* 10, no. 3 (2018): 333–360; and Nicholas Stern, "The Structure of Economic Modeling of the Potential Impacts of Climate Change: Grafting Gross Underestimation of Risk onto Already Narrow Science Models," *Journal of Economic Literature* 51, no. 3 (2013): 838–859.

17   IPCC, *Climate Change 2014*.

18   IPCC, 422.

19   Silke Beck and Martin Mahony, "The IPCC and the Politics of Anticipation," *Nature Climate Change* 7, no. 5 (2017): 311–313.

20   Larkin et al., "What If Technologies Fail?"; Detlef P. Van Vuuren et al., "Open Discussion of Negative Emissions Is Urgently Needed," *Nature Energy* 2 (2017): 902–904.

21   Simon Evans and Zeke Hausfather, "Q&A: How 'Integrated Assessment Models' Are Used to Study Climate Change," Carbon Brief, October 18, 2018, http://www.carbonbrief.org/qa-how-integrated-assessment-models-are-used-to-study-climate-change.

22   Massimo Tavon and Robert Socolow, "Modeling Meets Science and Technology: An Introduction to a Special Issue on Negative Emissions," *Climatic Change* 118, no. 1 (2013): 1–14; Detlef P. Van Vuuren et al., "Stabilizing Greenhouse Gas Concentrations at Low Levels: An Assessment of Reduction Strategies and Costs," *Climatic Change* 81, no. 2 (2007): 119–159.

23   Tavoni and Socolow, "Modeling Meets Science and Technology"; Beck and Mahony, "IPCC and the Politics of Anticipation."

24   Edward A. Parson, "Climate Policymakers and Assessments Must Get Serious about Climate Engineering," *Proceedings of the National Academy of Sciences* 114, no. 35 (2017): 9227–9230.

25   Leo Hickman, "Timeline: How BECCS Became Climate Change's 'Saviour' Technology," Carbon Brief, April 13, 2016, http://www.carbonbrief.org/beccs-the-story-of-climate-changes-saviour-technology.

26   Van Vuuren et al., "Stabilizing Greenhouse Gas Concentrations"; Christian Azar et al., "Carbon Capture and Storage from Fossil Fuels and Biomass—Costs and Potential Role in Stabilizing the Atmosphere," *Climatic Change* 74, nos. 1–3 (2006): 47–79.

27   Carton, "'Fixing' Climate Change."

28   Oliver Geden, "Politically Informed Advice for Climate Action," *Nature Geoscience* 11 (June 2018): 380–383; Oliver Geden, "The Paris Agreement and the Inherent Inconsistency of Climate Policymaking," *Wiley Interdisciplinary Reviews: Climate Change* 7, no. 6 (2016): 790–797.

29   Kate Dooley, Peter Christoff, and Kimberly A. Nicholas, "Co-producing Climate Policy and Negative Emissions: Trade-Offs for Sustainable Land-Use," *Global Sustainability* 1 (2018): 6.

30   Parson, "Climate Policymakers."

31   Dooley, Christoff, and Nicholas, "Co-producing Climate Policy," 7.

32   Evans and Hausfather, "Q&A."

33   Silke Beck and Martin Mahony, "The Politics of Anticipation: The IPCC and the Negative Emissions Technologies Experience," *Global Sustainability* 1 (2018): e8; Dooley, Christoff, and Nicholas, "Co-producing Climate Policy."

34   Esther Turnhout, "The Politics of Environmental Knowledge," *Conservation and Society* 16, no. 3 (2018): 363–371.

35   Esther Turnhout, Katja Neves, and Elisa De Lijster, "'Measurementality' in Biodiversity Governance: Knowledge, Transparency, and the Intergovernmental Science-Policy

Platform on Biodiversity and Ecosystem Services (IPBES)," *Environment and Planning A* 46, no. 3 (2014): 583.

36 Sean Low, "The Futures of Climate Engineering," *Earth's Future* 5 (2017): 67–71.

37 David Demeritt, "The Construction of Global Warming and the Politics of Science," *Annals of the Association of American Geographers* 91, no. 2 (2001): 307–337; Martin Mahony and Mike Hulme, "Epistemic Geographies of Climate Change," *Progress in Human Geography* 42, no. 3 (2016): 395–424; Beck and Mahony, "Politics of Anticipation."

38 Beck and Mahony, "Politics of Anticipation," 312.

39 Beck and Mahony, 5.

40 Beck and Mahony, 4.

41 Geden, "Paris Agreement."

42 Larkin et al., "What If Technologies Fail?"

43 Shue, "Climate Dreaming."

44 Evans and Hausfather, "Q&A"; IPCC, *Climate Change 2014*, chap. 6.

45 Cf. Van Vuuren et al., "Stabilizing Greenhouse Gas Concentrations"; and Keywan Riahi et al., "The Shared Socioeconomic Pathways and Their Energy, Land Use, and Greenhouse Gas Emissions Implications: An Overview," *Global Environmental Change* 42 (2017): 153–168.

46 Riahi et al., "Shared Socioeconomic Pathways."

47 IPCC, *Climate Change 2014*, sec. 6.3.1.3.

48 David Klein et al., "Global Economic Consequences of Deploying Bioenergy with Carbon Capture and Storage (BECCS)," *Environmental Research Letters* 11 (2016): 1–9.

49 IPCC, *Climate Change 2014*, chap. 6.

50 Elmar Kriegler et al., "Is Atmospheric Carbon Dioxide Removal a Game Changer for Climate Change Mitigation?," *Climatic Change* 118, no. 1 (2013): 45–57.

51 IPCC, *Climate Change 2014*, sec. 6.3.6.4.

52 Van Vuuren et al., "Stabilizing Greenhouse Gas Concentrations."

53 Van Vuuren et al., 131.

54 Kriegler et al., "Atmospheric Carbon Dioxide Removal."

55 Shell, "Sky Scenario," 23.

56 IPCC, *Climate Change 2014*; Van Vuuren et al., "Open Discussion."

57 Robert S. Pindyck, "The Use and Misuse of Models for Climate Policy," *Review of Environmental Economics and Policy* 11, no. 1 (2017): 100–114.

58 Elizabeth A. Stanton, Frank Ackerman, and Sivan Kartha, "Inside the Integrated Assessment Models: Four Issues in Climate Economics," *Climate and Development* 1, no. 2 (2009): 174.

59 Sheila Jasanoff, "A New Climate for Society," *Theory, Culture and Society* 27, no. 2 (2010): 242.

60 Larkin et al., "What If Technologies Fail?"; see also Beck and Mahony, "Politics of Anticipation."

61 Timothy E. Crews, Wim Carton, and Lennart Olsson, "Is the Future of Agriculture Perennial? Imperatives and Opportunities to Reinvent Agriculture by Shifting from Annual Monocultures to Perennial Polycultures," *Global Sustainability* 1 (2018): e11.

62 Karin Edstedt and Wim Carton, "The Benefits That (Only) Capital Can See? Resource Access and Degradation in Industrial Carbon Forestry, Lessons from the CDM in Uganda," *Geoforum* 97 (2018): 315–323; Sarah Milne et al., "Learning from 'Actually Existing' REDD+: A Synthesis of Etnographic Findings," *Conservation and Society* 17, no. 1 (2019): 84–95; Melissa Leach and Ian Scoones, eds., *Carbon Conflicts and Forest Landscapes in Africa* (New York: Routledge, 2015); Connor Cavanagh and Tor A. Benjaminsen, "Virtual Nature, Violent Accumulation: The 'Spectacular Failure' of Carbon Offsetting at a Ugandan National Park," *Geoforum* 56 (2014): 55–65; Esteve Corbera and Charlotte Friedli, "Planting Trees through the Clean Development Mechanism: A Critical Assessment," *Ephemera* 12, nos. 1–2 (2012): 206–241; Tracey Osborne, "Tradeoffs in Carbon

Commodification: A Political Ecology of Common Property Forest Governance," *Geoforum* 67 (2015): 64–77.

63 James Fairhead, Melissa Leach, and Ian Scoones, "Green Grabbing: A New Appropriation of Nature?," *Journal of Peasant Studies* 39, no. 2 (2012): 242.

64 Daniela F. Cusack et al., "An Interdisciplinary Assessment of Climate Engineering Strategies," *Frontiers in Ecology and the Environment* 12, no. 5 (2014): 280–287.

65 Van Vuuren et al., "Alternative Pathways."

66 Tavoni and Socolow, "Modeling Meets Science and Technology," 7.

67 Alyssa Battistoni, "There's No Time for Gradualism," *Jacobin*, October 9, 2018, http://jacobinmag.com/2018/10/climate-change-united-nations-report-nordhaus-nobel.

68 Kevin Anderson, "Duality in Climate Science," *Nature Geoscience* 8, no. 12 (2015): 898–900.

4

Defending a Failed Status Quo

─────────────────────●

The Case against
Geoengineering from a
Civil Society Perspective

LINDA SCHNEIDER AND LILI FUHR

In 2013, leading geoengineering advocate David Keith appeared on U.S. comedian Stephen Colbert's talk show. Having just published his book *A Case for Climate Engineering*,[1] it was an opportunity for Keith to introduce solar radiation management (SRM) to a giant viewership, even if he was sure to become the subject of Colbert's jokes: "Blanketing the Earth in sulfuric acid?" Colbert asked. "Is there any possible way that this could come back to bite us in the ass?" The audience roared with laughter.

Keith gamily played along: "Spray pollution into the atmosphere to stop it warming," he said, acknowledging the paradoxical nature of the theory behind solar geoengineering. Colbert responded, "So in the end, pollution saves them all." More laughter followed. "We owe pollution, we owe acid rain an apology, is what you are saying." Keith turned to underlining his supposedly reluctant, cautious approach, a rhetorical trademark of savvier geoengineering proponents: "It would be a totally imperfect fix. It would have risks. It wouldn't get us out of the need to stop polluting. But it might actually save people and be useful." And like any smart advocate, he gave a nod to one of the strongest arguments of his critics: "People are terrified of talking about this because they fear that it will prevent us from talking about cutting emissions," Keith admitted. Colbert joked, "Right, and also that it's sulfuric acid. . . . This is the all-chocolate dinner. I still get to have my CO_2 and I just have to spray sulfuric acid."[2]

On display were key aspects of the emerging public relations strategy of geoengineering advocates, but so too was a tremendously hopeful sign. The political sensibility behind Colbert's jokes—the belief that it is arrogant to presume to grasp the unintended consequences of technofixes and that it is folly to pursue such supposed solutions if they divert us from the task of reducing carbon emissions through reforms to our political and economic system—clearly resonated with the audience. Their laughter points to something that civil society can draw on: a deep reservoir of natural skepticism toward geoengineering among ordinary people, a visceral sense that earth systems should not be deliberately tampered with on a planetary scale. As geoengineering proposals gain more mainstream prominence in the coming years and its proponents seek to win public credibility and approval, the resistance of civil society will need to tap into and build off such instincts and common sense and channel them into political opposition.

Civil society around the world has long been fighting struggles against corporate power and control over our lives, the exploitation of humans and nature for the profit of a small minority, the violence of extractive and other transnational industries, and global social inequality and environmental destruction. The fight against polluting industries and the power they exert over our economies and politics is also at the core of the struggle for climate justice.

Geoengineering masquerades as an attempt to solve the climate crisis. In reality, it is yet another power grab. It would entrench existing inequalities, affirm existing power structures, and offer a means by which to avoid taking immediate climate action that might disrupt commercial interests.

As we argue in this chapter, geoengineering proposals serve the interest of high polluters; corporate, industrial, and military actors; and incumbent power structures. The links between geoengineering schemes and the fossil fuel and other industries, the potential financial profit to be gained from geoengineering technologies, and the risk of unilateral deployment, militarization, and violent conflict over impacts should be reason enough for a healthy skepticism toward geoengineering as a "solution" to climate change.

Why, then, consider such high-risk, planet-altering, extremely hubristic technologies at all? We don't think we should or have to. As long as we have not yet considered, let alone implemented, countless measures to drastically cut fossil fuel and industry emissions on more transformative socioeconomic pathways and the immediate protection and restoration of natural ecosystems,[3] there is no reason to buy into the inevitability of "hacking the planet."[4] Geoengineering, in turn, could never lead to a climate-just future and is fundamentally incompatible with core international principles and norms, such as the precautionary principle as well as vital planetary boundaries.

In this chapter, we further discuss international civil society's response to developments around geoengineering in several international fora, including the Hands Off Mother Earth (HOME) campaign, the strong support for the de facto moratorium on all climate-related geoengineering adopted by the U.N. Convention on Biological Diversity (CBD) in 2010, and current efforts to mobilize against the creeping normalization of geoengineering, specifically when it comes to policy and science debates in the Intergovernmental Panel on Climate Change (IPCC) and U.N. Framework Convention on Climate Change (UNFCCC).

Voices calling for investigation into geoengineering options are pointing to the inevitability of "negative emissions" and the need to prepare additional "insurance" tools against climate disaster in the form of solar radiation management. But these calls are also mobilizing many among civil society, academics, and government representatives of mostly Global South countries to point out the dangers of that path. This same broad spectrum postulates that the only viable, climate-just answer to the urgency of the climate crisis is a "radical realism" that recognizes the need for transformative climate action targeting the global inequality on which high-emission production patterns, fossil fuels, luxury emissions, and high-consumption lifestyles are built.

Opposition to and resistance against geoengineering can build on existing struggles and movements against the same industries and political elites that are benefitting from the status quo. Those in power have been for far too long delaying mitigation action. Now we can witness them investing in an effort to create powerful new—and partly military-controlled—industries that serve their interests and sustain the status quo. They seek to avoid stranding fossil fuel assets not by taking climate action but by subverting it. This move combines with a new geopolitics that further marginalizes those most impacted by climate change and with the least power to cope. The result is a diversion from investment in green, public, or community-owned and democratically controlled solutions. And the response we are beginning to see from civil society movements around the world is also clear and becoming louder: a no to technofixes and to new resource-intensive megainfrastructures and industries and a strong call to defend the right to have control over our territories, ecosystems, and food systems.

Defending the Incumbents: Shoring up Corporate and Political Power

Geoengineering schemes aim to intervene in the climate and natural ecosystems at a global scale. Integrated assessment models (IAMs) used in IPCC reports assume carbon dioxide removal (CDR) technologies will be able to extract several hundred to more than one thousand gigatons (Gt) of CO_2 from the atmosphere and store them underground or in the oceans. By way of comparison, current global CO_2 emissions per year stand at around 40 Gt—geoengineering claims it might be able to bury a staggering ten to thirty times that amount. SRM as a globe-spanning enterprise is equally vast in nature. Such planet-altering endeavors would use tremendous amounts of energy, land, water, rock material, and other resources. The sheer scale of geoengineering enterprises requires the establishment of new industries and helps sustain the incumbent polluting industries.

Five incumbent industries can be identified that stand to profit from geoengineering in particular: fossil fuels, agrochemicals, mining, the aerospace defense complex, and big data. Not by coincidence, these industries are among those that have stakes in the current economic status quo—a status quo that would be profoundly challenged if the world economy was to finally undergo the dramatic shift toward a socially just and ecologically sustainable future that climate change and other socioecological crises of the twenty-first century call for.

These are also the key incumbent industries that civil society has been mobilizing against for many decades because of the role they play in destroying and polluting nature, violating human rights, seizing power, and capturing the political space to make sure no regulation will hurt their profit interests. While their vested interests in the status quo are clear, civil society globally is only beginning to learn about their potential and real interests in new environmental modification technologies like geoengineering. Unveiling these links and interests will be a key stepping-stone in fueling civil society mobilization. A struggle against geoengineering thus becomes a part of and embedded in other important struggles against false solutions and for justice in the great transition.

The Fossil Fuel Industry

It is not by chance that the fossil fuel industry holds very favorable views of geoengineering, of CDR in particular but also increasingly of SRM, and first started conducting research into ways of removing CO_2 from the atmosphere decades ago.[5]

We are speaking about the same companies that for decades have invested in a multibillion-dollar deception campaign to cast doubt on the science of global warming and ensure that no meaningful political action was taken. What more could that industry wish for than a magical quick technofix that would (a) render transformational decarbonization strategies unnecessary and (b) provide a new excuse for maintaining a centrally controlled energy system and demand for new pipelines, extractive projects, and expertise in running large-scale industrial-military enterprises and infrastructures?

Carbon capture and storage (CCS), on which several CDR technologies rely, was originally developed by the oil industry as enhanced oil recovery (EOR), a technique to reach the final drops of oil from almost-depleted wells and reservoirs.[6] Both fossil CCS and CCS coupled with CO_2 sucked from the atmosphere with the help of direct air capture (DAC) and bioenergy plantations (for bioenergy with CCS [BECCS]) are now being rebranded as contributions to climate change mitigation, as Ina Möller explains in chapter 2. However, these proposed solutions are either not economically viable or unable to effectively remove CO_2 from the atmosphere: of the seventeen operational, commercial-scale CCS facilities worldwide that the Global CCS Institute lists in its 2017 report, thirteen use CO_2 for EOR. Of the four announced as coming on stream in 2018, three are for EOR.[7] This shows how the main interest in CCS and the only way to make it economically viable is oil production, which will increase rather than mitigate CO_2 emissions.

DAC suffers from a similar lack of effectiveness when it comes to removing CO_2 from the atmosphere. A number of DAC companies and start-ups currently run pilot-scale facilities that turn captured CO_2 into allegedly "CO_2-neutral" products that are more or less long-lived—these activities are usually referred to as carbon capture, usage, and storage (CCUS). Carbon Engineering, the commercial DAC company of Keith (who is, coincidentally, also one of the major proponents of SRM), aims to synthesize "clean transportation fuels" for use in the transportation, aviation, and shipping sector.[8] ClimeWorks in Switzerland, which operates another pilot DAC facility, claims to sell its CO_2 gas to "customers in key markets, including: food and beverage industries, commercial agriculture, the energy sector and the automotive industry."[9] In plain language, this means that the captured CO_2 is delivered to greenhouses or used for fizzy drinks or the energy, fuels, and materials markets, which includes use as feedstock for plastic production—another key area of operation of the petrochemical industry. While these commercial CCS and CCUS schemes link new and existing markets and business models and thereby entrench fossil fuel infrastructure and operations, they do nothing to address climate change.

In fact, due to life-cycle emissions arising from production, processing, and transportation, such CDR projects are, on balance, far from "carbon-neutral" and even less so "carbon-negative." Most CDR technologies produce so many new emissions at various stages in the production cycle that it appears highly unlikely that they could ever effectively remove large amounts of CO_2 from the atmosphere.

In the area of SRM, the most widely discussed technology—called stratospheric aerosol injection (SAI)—runs on large quantities of sulfur dioxide (SO_2), a major by-product of fossil fuel combustion. The continued reliance on BECCS, CCS, and other forms of geoengineering is enabling oil and gas majors to sustain wholly unrealistic and potentially catastrophic projections about the future demand for oil and gas products.

While many civil society groups engaged in the climate policy debate increasingly buy into the narrative that climate targets will not be met without "negative emission technologies" (NETs), there is also a strong potential within the broader climate movement to understand the underlying power dynamics and fossil fuel interests behind geoengineering and launch a vigorous campaign to prevent geoengineering from becoming the fossil fuel industry's latest maneuver of denialism and deception. It is also clear that the emerging CDR industries would inevitably operate on existing fossil fuel expertise and infrastructure, including a new wave of pipelines. Existing pipeline struggles could soon face a new frontier of pipeline construction for CDR industries and join the fight against geoengineering.

Agribusiness and Mining

The potential and real interest of the agribusiness and petrochemical sector in geoengineering becomes most visible when looking at BECCS: BECCS would require the conversion of vast amounts of land, including land currently used for food production, forests, and other natural ecosystems, into monoculture biomass plantations (fast-growing trees or crops). Chemical fertilizer inputs might double under such a scenario[10]—a boon for the already highly concentrated agrochemical industry. In a recent paper, Peters and Geden describe how the BECCS supply chain may well span several countries: "It could be that biomass harvested in Cameroon would be exported to the UK for combustion and CO_2 capture, and then the captured CO_2 exported to Norway for permanent storage."[11] Moreover, an elaborate carbon accounting system, harmonized across countries, would be needed to accurately measure, monitor, and verify the amount of carbon dioxide sequestered and stored.[12] While the authors' intention is to point out to policy makers that a transnational BECCS industry won't materialize all by itself but will require financial and political incentives and support, their description instructively prefigures certain characteristics of a potential future BECCS industry: the vast swaths of land required[13] would inevitably lead to an intensification of land privatization and land grabs in the Global South,[14] and experiences with the global expansion of biofuels—at a still much smaller scale—are the devastating precursors to what large-scale BECCS would hold in store. Processing, transportation of biomass, bioenergy production, CO_2 sequestration, and exportation to the site of final disposal would require cross-border and industrial-scale infrastructure, including heavy instruments used in industrial agriculture, processing plants, pipes and tubes, roads and railways, storage facilities, and so on . . . Who else would be better equipped to erect and maintain such industrial infrastructures than already-operating transnational industries, including those engaged in mining, fossil fuel production, and conventional agriculture (including fertilizer and pesticides)?

The expansion of agrofuels has already come with detrimental impacts for local communities and ecosystems, with large-scale land grabs, privatization, marginalization, loss of livelihoods and income, and the loss of biodiversity. BECCS at a climate-significant scale would imply such negative impacts on populations at a scale several orders of magnitude greater: it would lead to increasing competition over land, violations of land tenure rights, forced displacements (in particular in the Global South), surging food prices, and the undermining of livelihoods and income.

Large-scale BECCS would also require the transformation of natural ecosystems into monoculture tree or crop plantations and thereby bring about tremendous biodiversity loss. One study concluded that large-scale deployment of BECCS could result in a greater rate of terrestrial species extinction than a temperature increase of 2.8°C above preindustrial.[15] Moreover, turning diverse forests and other ecosystems into sealed-off monoculture "carbon farms" jeopardizes forest-dependent communities' livelihoods.

Furthermore, a dramatic increase in global fertilizer inputs through BECCS would exacerbate soil degradation and the anthropogenic disruption of the nitrogen cycle. Detrimental ecological impacts are a lack of oxygen supply in oceans, the eutrophication of streams and rivers, and a diminishing level of healthy nutrients in forests.[16]

Civil society organizations, including peasant movements and Indigenous peoples' organizations, have strong expertise and experience when it comes to opposing and resisting industrial agriculture, monoculture plantations, agrofuels, and pipelines. Movements like La Vía Campesina that represent millions of politically active peasants around the world have endorsed several statements against geoengineering and have begun educating their membership about its technologies. Their voices and power could come to fruition once BECCS moves from computer models to the real world.

Similar prospects arise from other CDR technologies, such as global-scale enhanced weathering, which involves distributing finely ground rock material (olivine or basalt) several millimeters thick onto agricultural fields (again, not just some agricultural fields but the largest part of the tropics). A study estimated that enhanced weathering could require olivine mining in the range of present-day global coal mining.[17] Given the long-standing and well-recognized track record of rights abuses and environmental destruction of extractive industries, geoengineering schemes are bound to further entrench industry power and marginalize local communities' rights and ecosystem integrity.

Global movements against large-scale mining and a broad range of civil society networks engaged in struggles against extractive industries have the expertise and political voice to make sure that no such ecologically outlandish ideas ever move from theory to practice.

Aerospace and Defense

The need for expansive resource-, land-, water-, energy-, and ultimately, emission-intensive industries to implement geoengineering schemes at a climate-relevant—that is, global—scale is the case not just for CDR but also for SRM. The most frequently discussed SRM technology is stratospheric aerosol injection (SAI), a scheme that envisages depositing large quantities of sulfur dioxide (SO_2) in the stratosphere to prevent a portion of the incoming sunlight from reaching the atmosphere. In a recent paper, Kravitz et al. calculate that toward the end of the twenty-first century, an annual rate of 51 teragrams (Tg; equivalent to 51,000,000 metric tons) of SO_2 would need to be introduced into the stratosphere to sustain the level of artificial cooling against the background of a high-emission scenario.[18] Beyond the staggering quantities of material involved in solar geoengineering schemes, SRM also involves large-scale aerial infrastructures to carry out the particle injection at frequent intervals over an extended period of time and at various sites around the globe. It is evident from the magnitude and degree of sophistication of the required apparatus that powerful states with highly developed air fleets would be best equipped to implement a potential SAI deployment. Such endeavors are unlikely to be feasible without the involvement of national or transnational aerospace and defense industries and would open up new fields of operation as well as a possibility to further deepen power and social control in the name of climate protection.

Big Data

Geoengineering schemes would quite evidently be in the interest of the fossil fuel and other extractive industries in that they rely on their infrastructure, services, expertise, and promise of the possibility of cleaning up after the act. Yet they also promise expanded application, commercial gains, and greater power for actors developing other new and emerging technologies such as big data.

Virtually any geoengineering scheme, for technical or political and governance reasons, would require a globe-spanning grid of measuring sites to monitor and control climatic and weather parameters. This is especially true for SRM schemes such as SAI and marine cloud brightening that rely on constantly recharging the climate with large quantities of substances to uphold the artificial cooling and to avoid triggering a "termination shock" that would unleash a sudden uptick in temperature, with rates of climatic change well beyond what many species (including, arguably, humans) could adapt to.[19] But permanent monitoring, reporting, and verification for carbon sequestration and storage in soils, in biomass, in the oceans, and underground would equally demand the extraction of massive amounts of data. Geoengineering deployment and its perpetuation over centuries are infeasible without the constant universal surveillance of the climate and other earth systems.

But geoengineering also draws on and helps proliferate other planet-altering, disruptive technologies controlled and owned by transnational corporations, including genetic engineering: for instance, fast-growing, genetically modified trees large may be used for afforestation schemes. Moreover, technologies that aim to "enhance" the photosynthetic or reflective properties of crops rely on synthetic biology,[20] as do many algae schemes that involve carbon-sequestering microorganisms.

Finally, even the emerging artificial intelligence market has a potential stake in geoengineering. In the same solar geoengineering study by Kravitz et al., a specialized algorithm was developed for SRM deployment simulations that would, incorporating feedback from the climate system, continually adjust the optimal amount and locations of sulfur dioxide dispersal. The authors found that by automatizing the choice of dosage and injection site, such artificial intelligence-driven, fine-tuned SAI deployment led to more even and uniform modeling results than when human researchers were setting those same parameters.[21] Full automation of geoengineering deployment using drones and other remote-controlled devices would be a logical next step.

With technological systems of geoengineering, artificial intelligence, synthetic biology, and big data converging, democratic governance of technology development will recede further into the distance while shoring up corporate interests and deepening the penetration of corporate control over human and natural processes.

In light of these dystopian visions, it becomes clear: the geoengineering debate would certainly profit from the expertise and rich knowledge of those actors within civil society working to protect and reclaim fundamental human rights in the digital age. The digital rights movement could protest surveillance and control (be it corporate or government driven) and experiment with and envision emancipatory, citizen-driven technologies and data for the common good.

. . . While Making Matters Worse

From the setup and shape of the geoengineering discourse over the past decade—dominated by white, middle-aged men from highly industrialized countries with scientific backgrounds in, most typically, physics or engineering and undergirded by a political economy of fossil fuel and other powerful industries as well as military interests—we can infer the real-world conditions under which geoengineering schemes would unfold.

In the context of the actual political economy in which geoengineering emerges, chances are high that northern, capital-backed voices will dominate decision-making over geoengineering deployment, and the interests of the vast majority of people who would suffer most from the impacts of geoengineering will be easily sidelined. Unilateral deployment of SRM by a powerful state or a coalition of states, or its weaponization, should not be underestimated.[22]

While there is no doubt about the inexcusable inaction that has characterized much of the past decades, especially in the high-emitting Global North, the way forward must not further entrench existing global inequalities or power structures or exacerbate concurrent global socioecological crises. This is true for SRM and CDR, the latter of which is not at all necessarily more benign.

Moreover, despite claims to the contrary, geoengineering would not only not solve the climate crisis; it would in fact make matters much worse—for the climate and for human communities and ecosystems that are already heavily strained and will disproportionally suffer from the impacts of climate change. Given the political economy context in which geoengineering schemes are currently developed, there is a great risk that they would exacerbate concurrent global socioecological crises such as biodiversity loss, degradation of natural ecosystems, land grabs and soil loss, social inequality and human rights abuses, poverty, marginalization, and hunger.

The expansion of agrofuels has already come with detrimental impacts for local communities and ecosystems, with large-scale land grabs, privatization, marginalization, the loss of livelihoods and income, and the loss of biodiversity.

In the context of marine geoengineering, one of the most frequently discussed technologies is ocean fertilization, which involves dumping large quantities of iron or other nutrients to enhance the growth of phytoplankton in marine areas with lower primary productivity. The additional plankton would sequester CO_2 from the atmosphere and eventually die and sink to the ocean floor, where, so the theory goes, the carbon would remain "stored." Around one dozen field experiments have already been carried out in the past and have proven ocean fertilization to be virtually ineffective. The vast majority of additional plankton was ingested by other marine organisms and ended up in the marine food web, failing to remove the CO_2 permanently from the cycle. Furthermore, only a negligible fraction actually settled down as ocean sediment. However, being ineffective as a means to remove carbon dioxide from the atmosphere does not mean a geoengineering intervention is harmless or without effect on the ecosystem it targets: ocean fertilization intervenes heavily in the ocean chemistry and changes the structure and composition of a marine ecosystem's populations. The massive growth of algae can deplete the supply of oxygen near the surface and cause a mass extinction of fish. Toxic algal blooms may be another unintended side effect. All these would threaten the livelihoods of fisher folks and coastal communities that depend on intact marine ecosystems. And yet new outdoor experiments are currently underway.[23]

At an ontological level, geoengineering relies on a problem description that is purely technical: it reduces climate change to a matter of CO_2 molecules in a certain place rather than another and a unit of radiative forcing—two technical components that could be manipulated and thereby rendered inoperative. This is a deeply misleading approach to both the socioeconomic causes of climate change and the sociopolitical ecology of its impacts and fails to take into account the complexity and interconnectedness of local, regional, and global ecosystems.

Geoengineering's reductionist way of dissecting the world into carbon-storing and sunlight-reflecting entities is hostile to a holistic, integrated appreciation of ecological systems and therefore is blind to the severe damages it could inflict on ecosystems and human communities. It commits an extreme form of "ecological epistemicide."[24] The answer to why such approaches are nonetheless promoted and pushed into the mainstream of climate policy is in no small part to be found in what has been explored in part 2.

Radical Realism for the Transformation We Need

Civil society does have the potential to not only engage much more in struggles against incumbent industries and political elites that would profit from geoengineering but also greatly increase its fight for transformative and radical emission reduction strategies that are urgently needed.

The alleged inevitability of large-scale technological manipulation of natural processes and ecosystems to address devastating climate change is usually fervently argued with recourse to climate-economic models—so-called IAMs—that are almost always conservative in the sense of preserving the status quo, as Wim Carton explains in chapter 3. They presuppose neoclassical economic assumptions, including continued global economic growth and ever-rising consumption; prioritize technological solutions, simplistic cost-benefit analyses, and shifting the burden onto the next generation; largely exclude broader societal transformations and changes in the underlying economic system; and not least of all, project an image of an apolitical, "neutral," and scientific account of the options to respond to climate change.

However, what is consistently and systematically neglected and downplayed in mainstream climate policy debates is the potential for much deeper emissions reductions if we can break with incumbent power structures. Among the options that have not even been considered, let alone implemented as mitigation options, are an early phaseout not just of global coal production but also of oil and gas, including the early retirement of existing and producing infrastructure; a shift to 100 percent decentralized renewable energy production and supply from solar and wind;[25] a global transition toward an agro-ecological smallholder agriculture that would save vast amounts of emissions from conventional industrial agriculture and also allow for food sovereignty;[26] the implementation of circular economy approaches to reduce the absolute amount of energy and resources metabolized in the global economy; and an economic model that does not rely on endless growth.[27] With the attendant redistribution of global wealth and income between and within countries and drastically reduced socioeconomic inequality, such approaches would both be affordable and not further hurt marginalized communities. Such climate-just solutions to the climate crisis—which is a socioecological crisis, not an engineering problem—are feasible and viable and would render geoengineering schemes obsolete. Yet incumbent power structures and vested interests prevent us from implementing them.

Beyond the potential for much deeper and much more rapid emissions reductions, there are also ways of drawing excess CO_2 from the atmosphere that are more safe and more socially as well as ecologically sustainable than industrial-scale CDR technologies. These include the ecological restoration of degraded and already destroyed ecosystems—above all, forests but also moors, peatland, and marine ecosystems, among others. Their protection and careful regeneration would best be implemented by safeguarding the communal land tenure rights of those who have for centuries treated these ecosystems in an ecologically sustainable way—Indigenous and other local communities—rather than privatizing them.[28] Quite conspicuously, such natural climate solutions do not receive the kind of attention from climate scientists and climate policy experts that is granted to CDR technologies despite the fact that they are proven, safe, ready to implement, and are also capable of protecting human and land rights as well as ecosystem integrity.

Civil Society Struggles around Geoengineering Governance

None of the proposed geoengineering technologies currently exist at scale. While some are currently being tested in pilot projects, many others remain fraught with fundamental uncertainty if they could ever work at all. Given the state of development of most geoengineering schemes, it is reasonable enough that they appear as far-fetched chimera to most people. However, the push for geoengineering to enter the mainstream of international climate policy should not be underestimated. Already, CDR technologies are taken for granted in climate-economic models and virtually all mainstream climate policy discussions. The IPCC has decided to make geoengineering a crosscutting issue in its upcoming sixth assessment report (AR6) and discusses both CDR and SRM in its influential special report *Global Warming of 1.5°C*, published in the fall of 2018.[29] Several outdoor experiments have been proposed for the near future that would try to prepare the ground for more large-scale geoengineering testing. The actors promoting these experiments are exploiting the absence of a global and robust governance mechanism that regulates not only potential deployment but also outdoor experimentation—including the possibility of prohibiting both. They ignore or downplay the de facto moratorium put in place by the CBD, arguing that this is a matter that should be discussed in the climate context. And while geoengineering has so far been only a much talked about issue in the corridors of the UNFCCC, it might enter the negotiation process formally through the Talanoa Dialogue as early as this year.

Active resistance to geoengineering has so far been limited to engagement in a few key international fora dealing with governance, on the one hand, and protests around outdoor experiments in the case of ocean fertilization and solar radiation management, on the other. It is, however, important to note that this is changing quickly, with geoengineering moving up the political agenda after the ratification of the Paris Agreement with its ambitious climate targets and vastly insufficient national pledges to reduce greenhouse gas (GHG) emissions.

There is an additional growing momentum to bring the idea of "natural climate solutions" to the forefront of the climate policy agenda, pointing to the fact that the only reliable and proven way of pulling excess carbon out of the atmosphere remains to protect natural ecosystems and carefully restore degraded ones.[30]

In what follows, we will give a brief outline of the first global campaign against geoengineering as well as the discussions in different U.N. fora like the CBD and the London Convention, tracing how the moratoria in these key fora were achieved. We will then describe how protests against outdoor experiments have led to the cancelation or severe delay of those experiments. We will explore resistance against geoengineering as it shifts to a fight in defense of Indigenous rights and territories and finish by exploring current efforts to mobilize against the "normalization" of geoengineering.

We observe that we are right now witnessing the emergence of a new international alliance, perhaps even a movement, against geoengineering that has the power to mobilize the silent majority against the few loud and aggressive pro-geoengineering voices that currently dominate the conversation.

Hands Off Mother Earth

The first proactive civil society campaign against geoengineering was launched in 2010 in Bolivia at the World People's Conference on Climate Change and the Rights of Mother Earth, which gathered more than thirty-five thousand participants, mostly Indigenous. The conference-wide Peoples' Agreement included an explicit denunciation of geoengineering, but the conference was also the site of a new campaign's launch. Hands Off Mother Earth (HOME) was supported by over one hundred organizations, including the Indigenous Environmental Network, Acción Ecológica of Ecuador, Friends of the Earth International, the Third World Network, and La Vía Campesina.

The ETC Group, a small technology watchdog organization with offices in Canada, Mexico, the Philippines, and the United Kingdom, was a crucial leader of the effort, as it has been for much of the monitoring, education, and campaigning around geoengineering. Signatories to the campaign included prominent individual signatories like David Suzuki, Bill McKibben, Naomi Klein, and Vandana Shiva. The campaign described itself as a "clear message to the geoengineers and to governments worldwide that our home is not a laboratory." The campaign's website created a portal where individuals could submit photos of their hand, acting as an endorsement of their call to keep "Hands Off Mother Earth."[31]

It was not the first global statement critical of geoengineering. Scientists for Global Responsibility had published the first warning in a report in 1996, and Greenpeace U.K. had followed up with its own report in 1999. Ten years later, the Indigenous Peoples' Global Summit on Climate Change in Alaska brought together four hundred Indigenous people from eighty countries and issued the "Anchorage Statement," calling geoengineering a "false solution." That same year, in 2009, seventy civil society organizations had signed a statement opposing the privately organized Asilomar Geoengineering Conference.

Besides a goal of public education, the HOME campaign's main objective, launched in April 2010, was to build momentum toward achieving a moratorium on geoengineering at the meeting of the CBD to be held later that year.

Geoengineering Discussions at the United Nations

The United Nations has been home to a decade-long discussion on geoengineering based on the precautionary approach and environmental and social concerns, with its center of gravity at the CBD. Those debates were driven by civil society, academics, and governments from developing countries concerned about the risks and dangers of geoengineering. Various decisions in the United Nations can be traced back to being in direct response to geoengineering plans or actual deployment of outdoor experiments. It is important to note that despite the clear links to climate change concerns, these debates in the CBD were not (and until today are still not) reflected in the UNFCCC context. Climate change negotiators and experts, unfortunately, remain largely ignorant when it comes to expertise and decisions of the CBD that relate to climate change and ecosystem integrity.

At the CBD, a de facto moratorium on ocean fertilization was established as early as 2008. In late 2010, on the heels of growing pressure built by the HOME campaign, parties to the CBD agreed to a de facto international moratorium on all climate-related geoengineering. The CBD has published two reports on geoengineering that were extensively reviewed by its member governments, including an analysis of the regulatory and legal framework related to the convention and the possible role of other U.N. bodies.

More thematically focused and with recourse to the precautionary principle and expected adverse impacts on the marine environment, the London Protocol of the London Convention to prevent marine pollution adopted a decision in 2013 to regulate marine geoengineering. Its annex 4, which lists prohibited marine geoengineering activities, for now only contains ocean fertilization—but the mechanism is deliberately construed with the flexibility to include further technologies of marine geoengineering.

Climate manipulation has also been a subject of military interests for many decades as a means to control the weather for hostile purposes. The impacts of the hostile use of weather modification by the United States against Vietnam led to the adoption of the U.N. Environmental Modification Treaty (ENMOD) in 1977 to prevent the manipulation of the environment as a means of warfare. ENMOD entered into force in October 1978 and thus saw its fortieth anniversary in 2018. In light of technological advances in the fields of synthetic biology and geoengineering with clear interests of military actors and concrete plans to release substances and organisms that have the potential to alter entire ecosystems at a planetary scale, we can expect increased interest to call on ENMOD to discuss these matters in the coming years.

Geoengineering proponents are largely and intentionally denying the reality of these discussions that have already taken place inside the U.N. system. Many also disclaim any military component or relevance of this debate, ignoring the fact that some of the research is Pentagon-funded. They argued instead that geoengineering research and experiments can be self-regulated and voluntarily managed through ethical guidelines, codes of conduct, and similar measures. Some pragmatically believe that such "soft governance" approaches are more in line with the way international governance moves forward in the current geopolitical climate, while others are hoping that some kind of self-regulation or soft regulation of the first links in the geoengineering chain would prevent broader international measures, such as a ban.

Protests against Outdoor Experiments

Outdoor experiments of geoengineering have always triggered intense debates and local protests and have thus been the key in advancing criticism of the technology. Here are two key examples:

The "Spice Boys"

From 2010 to 2012, the United Kingdom was host to the first attempted outdoor experiment advancing stratospheric aerosol injection, which invited public attention and debate and then was canceled before, literally, getting off the ground. Known as the stratospheric particle injection for climate engineering—or SPICE and nicknamed "Spice Boys" by its detractors—it was designed to test hardware for a larger-scale deployment of the technology. The idea for SPICE was born in a "sandpit"—a short, cross-disciplinary meeting intended to generate innovative ideas—that was run by three of the United Kingdom's seven research councils. Involving climate modelers, chemists, and engineers, the project was backed by four universities, several government departments, and the private company Marshall Aerospace.

The experiment was to test a kilometer-long hose suspended by a giant helium balloon. A pump would deliver a few dozen liters of water to the top of the hose, where it would be sprayed as a mist, evaporating before it hit the ground. It was scheduled to take place on a disused military airstrip in Norfolk, United Kingdom, in the fall of 2011. While the experiment would likely have had no discernible effect on the environment, the ETC Group called it a Trojan hose that would dangerously open the door for large-scale deployment of solar radiation management.

Soon after the press conference launch, the public backlash began. An open letter signed by fifty organizations from around the world asked the U.K. Government and Research Councils to scrap the experiment. Press controversy ensued. Within a week, the researchers and the research council backing them decided to delay the experiment. The SPICE researchers told the media it was delayed not because of the public outrage but because of a conflict of interest: they had discovered two scientists involved had not disclosed that they had submitted patents for similar technologies before the experiment was proposed. Yet it seems implausible that the public criticism was irrelevant to the decision. An additional concern was no doubt that the negative reactions to SPICE would threaten subsequent research and experimentation on geoengineering. By April 2012, the experiment was permanently canceled.

Russ George, the Ocean Fertilization "Geo-vigilante"

The most persistent advocate of ocean fertilization, a marine-based geoengineering approach that has been banned under the CBD and London Protocol to the London Convention (see earlier), has been U.S. businessman Russ George. More than ten years ago, he created a U.S. start-up company, Planktos, that by early 2007 was selling carbon offsets on its website. Planktos claimed that its initial ocean fertilization test, conducted off the coast of Hawai'i from singer Neil Young's private yacht, was taking carbon out of the atmosphere. Soon thereafter,

Planktos announced plans to set sail from Florida to dump tens of thousands of pounds of iron particles over ten thousand square kilometers of international waters near the Galápagos Islands, a location chosen because, among other reasons, no government permit or oversight would be required.

Several groups, including Acción Ecológica, ETC Group, Institut de Ciència i Tecnologia Ambientals (ICTA-UAB), Greenpeace, International Union for Conservation of Nature (IUCN), Indigenous Environmental Network (IEN), Carbon Trade Watch, and the Durban Group took part in mounting a critical response. The Sea Shepherd, a marine conservation organization, was reported to be sending a ship to "intercept" the Planktos ship. Hearing this, the Planktos ship changed course and set off to Spain and the Canary Islands. Hit with negative publicity, Planktos announced it was indefinitely postponing its plans due to a "highly effective disinformation campaign waged by anti-offset crusaders." In April 2008, Planktos declared bankruptcy, sold its vessel, dismissed all employees, and claimed it had "decided to abandon any future ocean fertilization efforts."

That was not to be. George reappeared a few years later, having persuaded a band council of the Indigenous Haida nation on the archipelago of Haida Gwaii to fund a new project. This time, incorporated as the Haida Salmon Restoration Corporation (HSRC), he pitched iron fertilization as a way to boost salmon populations, with the added benefit of selling carbon credits based on sequestering CO_2 in the ocean. In 2012, news broke that he had orchestrated a dump of one hundred tons of iron sulfate in the Pacific Ocean off the west coast of Canada—the largest ever ocean fertilization dump. Critical media coverage in the *Guardian*, the Canadian Broadcasting Corporation, the *New York Times*, and elsewhere ensued, showing again that real-world experiments provide crucial opportunities for opponents of geoengineering to reach a broader public readily sympathetic to their message. An international outcry landed George with the mantle of a "rogue geoengineer" and "geovigilante" and made him the target of an investigation by Environment Canada's enforcement branch (which, eight years later, has strangely yet to conclude or disclose its findings).

Many of those involved in this Haida project have again resurfaced, this time as the Vancouver-based Oceaneos Marine Research Foundation. Their sights now are on an experiment off the shores of Chile, where they say they are seeking permits from the Chilean government to release up to ten metric tons of iron particles. They have rebranded, presenting their organization as nonprofit rather than for-profit, as engaging in "ocean seeding" rather than iron fertilization, and as scrupulously following a code of conduct and a board of scientific advisors. The project has been sharply criticized by ocean scientists in Chilean research institutions.[32]

In Defense of Rights and Territories— Geoengineering and Indigenous Resistance

As an approach that treats the living globe as an "engineerable" subject and whose strongest proponents are transnational corporate actors and mostly Western natural scientists and engineers, it is no surprise that some of the most trenchant critiques of geoengineering have come from Indigenous peoples, who espouse a place-based and more sacred relationship with mother earth, and their movements.

The very idea of geoengineering recasts the global climate and other earth systems as mechanistic processes that can be pragmatically altered with a Herculean science project. Indigenous movements worldwide are increasingly setting the front lines of resistance to fossil fuel and extractive projects, citing the rights of mother earth and sacred defense of land and water. Resisting geoengineering is emerging as part of that struggle. This is particularly relevant given that the outdoor experiments currently planned in North America will happen on or affect Indigenous lands, territories, and rights.

Future Frontlines

The history of resistance against geoengineering and the history of governance debates in key U.N. fora described previously clearly show that this is neither a new topic nor one that concerns just a handful of nongovernmental organization (NGO) representatives, as many of the geoengineers like to claim, downplaying the scale and depth of opposition spanning more than a decade.

Key current frontlines of resistance in the geoengineering debate, while widespread, are not yet well connected enough to allow for a movement to become visible across issues, geographical barriers, and constituencies. This remains a key task for the immediate future.

We expect Indigenous peoples and environmental and climate justice groups in North America to take a leading role in bringing together this movement of movements—ranging from farmers to fishers, from NGO activists to social movement leaders, from critical scientists to progressive and bold government representatives.

Civil society is beginning to coordinate and collaborate more closely with impacted communities. Several groups are preparing to relaunch the HOME campaign. In the United States, the Climate Justice Alliance and the Indigenous Environmental Network, alongside the ETC Group and others, are beginning to ground the fight against geoengineering as a defense of Indigenous territory, responding to proposed geoengineering experiments in the United States. These include a stratospheric aerosol injection experiment by the Harvard Solar Geoengineering Program,[33] led by Keith, planned to be carried out in Arizona near the U.S.-Mexico Border; a marine cloud brightening experiment planned for Monterey Bay, California;[34] and the Ice911 project, which aims to spread glass microbeads over sea ice in Alaska.[35] Ongoing efforts by the Oceaneos ocean fertilization project in Chile and Peru are also sure to spark protests in South America.[36]

Plans of the U.S. administration to work toward a strategic research program on geoengineering while handing out massive tax cuts to CCS-based technologies may also incite a new level of resistance and collaboration.[37]

Since many proponents of geoengineering vigorously deny any connection to the commercial interests of the fossil fuel industry but early evidence demonstrates persistent and pervasive links between oil, gas, and geoengineering, further uncovering those links will likely strengthen the opposition and weaken the proponents. Another crucial obstacle to overcome in order to mobilize greater parts of the environmental movement and public will be to make inroads among climate organizations. For instance, 350.org, which has helped turn the mainstream U.S. environmental movement toward more grassroots, direct action–oriented campaigns and mounted large-scale actions to keep fossil fuels in the ground, could be a powerful ally in campaigns against false solutions like geoengineering. While McKibben, one of the founders of 350.org, has expressed critical attitudes toward geoengineering, the organization, with chapters in countries across the world, has yet to take a formal stance.

As geoengineering increasingly emerges and becomes visible as the new form of denialism by the fossil fuel industry and its allies in government, the growing global movements working to keep oil, gas, and coal in the ground and working to undermine the political and economic power of the fossil fuel industry could join the geoengineering resistance. A grand coalition of the climate movement, the peasants and food sovereignty movement, Indigenous peoples, antiwar activists, digital rights activists, and others would be a powerful force indeed.

We, therefore, are optimistic that civil society and the global public will not so easily buy into the geoengineering narrative. Instead, we think they will react aversely to its proponents and remain hopeful and realistic that radical emission reductions, as part of much deeper, transformative socioeconomic and socioecological pathways, as well as ecosystem integrity and natural climate solutions will be the best hope for dealing with the planetary climate crisis. Geoengineering is only fuel to the fire—a desperate attempt to uphold a long-failed status quo.

Rejected as the false solution that it is, we will be able to rest assured that geoengineering advocates will not be able to tamper with planetary systems. Instead, we will be able to laugh with some relief at the comically dangerous image of Keith evoked by Colbert—"a man in a hollowed-out volcano with henchmen, who occasionally shakes his fist at the sky and says 'they said I was a fool at Harvard. Who's the fool now?!'"[38]

Notes

1   David Keith, *A Case for Climate Engineering* (Cambridge, Mass.: MIT Press, 2013).
2   Patrick Kevin Day, "Stephen Colbert's Guest Says Sulfuric Acid Could Stop Global Warming," *Los Angeles Times*, December 10, 2013, http://notices.californiatimes.com/gdpr/latimes.com/.
3   Heinrich Böll Foundation, *Radical Realism for Climate Justice: A Civil Society Response to the Challenge of Limiting Global Warming to 1.5°C* (Berlin: Heinrich Böll Foundation, 2018).
4   Raymond T. Pierrehumbert, "The Trouble with Geoengineers 'Hacking the Planet,'" *Bulletin of the Atomic Scientist*, June 23, 2017.
5   Center for International Environmental Law (CIEL), "Fuel to the Fire: How Geoengineering Threatens to Entrench Fossil Fuels and Accelerate the Climate Crisis," press release, February 13, 2019, http://www.ciel.org/news/fuel-to-the-fire-how-geoengineering-threatens-to-entrench-fossil-fuels-and-accelerate-the-climate-crisis/.
6   International Energy Agency (IEA), *Storing CO$_2$ through Enhanced Oil Recovery: Combining EOR with CO$_2$ Storage (EOR+) for Profit* (Paris: IEA, 2015).
7   Global CCS Institute, *The Global Status of CCS* (Melbourne: Global CCS Institute, 2017).

8   "Air to Fuels," Carbon Engineering, accessed May 15, 2019, http://carbonengineering
    .com/about-a2f/.

9   "Our Customers," Climeworks, accessed December 15, 2017, http://www.climeworks
    .com/our-customers/.

10 Felix Creutzig, "Economic and Ecological Views on Climate Change Mitigation with
    Bioenergy and Negative Emissions," *Global Change Biology Bioenergy* 8, no. 1 (2014): 4–10.

11 Glen P. Peters and Oliver Geden, "Catalysing a Political Shift from Low to Negative
    Carbon," *Nature Climate Change* 7 (2017): 619–621.

12 Such accounting would have to accurately capture the additional CO_2 arising in bioenergy
    production from fertilizer inputs, transportation, processing, and extra energy required for
    CCS—all these are often readily swept under the carpet to maintain the belief in bioener-
    gy's purported carbon neutrality. For a growing body of literature refuting this fundamental
    flaw underlying the push for bioenergy, see Mary S. Booth, "Not Carbon Neutral: Assessing
    the Net Emissions Impact of Residues Burned for Bioenergy," *Environmental Research
    Letters* 13 (2018): 035001; and John D. Sterman, Lori Siegel, and Juliette N. Rooney-Varga,
    "Does Replacing Coal with Wood Lower CO_2 Emissions? Dynamic Lifecycle Analysis of
    Wood Bioenergy," *Environmental Research Letters* 13, no. 1 (2018): 015007.

13 Estimates vary and, by nature, depend on the amount of BECCS that a given model
    assumes but range from several million hectares to over a billion hectares of productive
    land, according to Christopher B. Field and Katharine J. Mach, "Rightsizing Carbon
    Dioxide Removal," *Science* 356, no. 6339 (2017): 706–707. By way of comparison, total
    global cropland encompasses 1.5 billion hectares.

14 This is true for most of the terrestrial CDR technologies and even afforestation schemes,
    which are usually deemed relatively benign: more often than not, afforestation implies
    monoculture tree plantations on privatized land that may displace diverse forests, other
    ecosystems, and people depending on those ecosystems for sustenance, livelihood, and
    habitat.

15 Phil Williamson, "Emissions Reduction: Scrutinize CO_2 Removal Methods," *Nature* 530
    (2016): 153–155.

16 Sivan Kartha and Kate Dooley, "Land-Based Negative Emissions: Risks for Climate Miti-
    gation and Impacts on Sustainable Development," *International Environmental Agree-
    ments: Politics, Law and Economics* 18, no. 1 (2018): 79–98; for an overview of risks and
    concerns around BECCS, see also Wil Burns and Simon Nicholson, "Governing Climate
    Engineering," in *New Earth Politics: Essays from the Anthropocene*, ed. Simon Nicholson
    and Sikina Jinnah (Cambridge, Mass.: MIT Press, 2016), 343–366.

17 Peter Köhler, Jens Hartmann, and Dieter A. Wolf-Gladrow, "The Geoengineering Poten-
    tial of Artificially Enhanced Silicate Weathering of Olivine," *Proceedings of the National
    Academy of Sciences of the United States of America* 107 (2010): 20228–20233.

18 Ben Kravitz et al., "First Simulations of Designing Stratospheric Sulfate Aerosol Geo-
    engineering to Meet Multiple Simultaneous Climate Objectives," *Journal of Geophysical
    Research: Atmospheres* 122 (2017): 12616–12634.

19 Christopher H. Trisos et al., "Potentially Dangerous Consequences for Biodiversity of
    Solar Geoengineering Implementation and Termination," *Nature Ecology & Evolution* 2
    (2018): 475–482.

20 ETC Group and Heinrich Böll Foundation, *Outsmarting Nature? Synthetic Biology and
    Climate Smart Agriculture* (Val David, Canada: ETC Group; Berlin: Heinrich Böll Foun-
    dation, 2015).

21 Kravitz et al., "First Simulations."

22 Andy Parker, Joshua Horton, and David Keith, "Stopping Solar Geoengineering through
    Technical Means: A Preliminary Assessment of Counter-Geoengineering," *Earth's
    Future* 6 (2018): 1058–1065.

23 See the interactive world map at http://map.geoengineeringmonitor.org/ for an overview
    of past, ongoing, and planned geoengineering experiments and research projects.

24 Camila Moreno, Daniel Speich Chassé, and Lili Fuhr, *Carbon Metrics: Global Abstractions and Ecological Epistemicide*, 2nd ed. (Berlin: Heinrich Böll Foundation, 2016).

25 Mark Z. Jacobson et al., "100% Clean and Renewable Wind, Water, and Sunlight All-Sector Energy Roadmaps for 139 Countries of the World," *Joule* 1 (2017): 108–121.

26 ETC Group, *Who Will Feed Us? The Peasant Food Web vs. the Industrial Food Chain*, 3rd ed. (Val-David, Canada: ETC Group, 2017).

27 See Heinrich Böll Foundation, *Radical Realism for Climate Justice: A Civil Society Response to the Challenge of Limiting Global Warming to 1.5°C* (Berlin: Heinrich Böll Foundation, 2018).

28 Kate Dooley et al., *Missing Pathways to 1.5°C: The Role of the Land Sector in Ambitious Climate Action* (n.p.: Climate Land Ambition and Rights Alliance [CLARA], 2018).

29 Intergovernmental Panel on Climate Change (IPCC), *Global Warming of 1.5°C: An IPCC Special Report on the Impacts of Global Warming of 1.5°C above Pre-industrial Levels and Related Global Greenhouse Gas Emission Pathways, in the Context of Strengthening the Global Response to the Threat of Climate Change, Sustainable Development, and Efforts to Eradicate Poverty* (Geneva, Switzerland: IPCC, 2018).

30 Dooley et al., *Missing Pathways*.

31 Hands Off Mother (HOME; website), accessed April 15, 2019, http://www.handsoffmotherearth.org.

32 Jeff Tollefson, "Plankton-Boosting Project in Chile Sparks Controversy," *Nature* 545 (2017): 393–394.

33 Arthur Nelsen, "US Scientists Launch World's Biggest Solar Geoengineering Study," *Guardian*, March 24, 2017, http://www.theguardian.com/environment/2017/mar/24/us-scientists-launch-worlds-biggest-solar-geoengineering-study.

34 "Marine Cloud Brightening Experiment Briefing," Geoengineering Monitor, April 6, 2018, http://www.geoengineeringmonitor.org/2018/04/marine-cloud-brightening-project-geoengineering-experiment-briefing/.

35 "Ice911 Experiment Briefing," Geoengineering Monitor, April 6, 2018, http://www.geoengineeringmonitor.org/2018/04/ice-911-geoengineering-experiment-briefing/.

36 Silvia Ribeiro, "Pirates of the Pacific," Geoengineering Monitor, July 12, 2018, http://www.geoengineeringmonitor.org/2018/07/pirates-of-the-pacific/.

37 See the reformed 45Q Tax Credit of February 2018 as well as the Geoengineering Research Evaluation Act that was introduced in Congress in December 2017.

38 "David Keith," *Colbert Report*, Comedy Central, Season 10, Episode 33, September 12, 2013, http://www.cc.com/video-clips/lvohd2/the-colbert-report-david-keith.

5

Geoengineering and
Indigenous Climate Justice

────────────────────●

A Conversation with
Kyle Powys Whyte

KYLE POWYS WHYTE, INTERVIEWED
BY HOLLY JEAN BUCK

Kyle Powys Whyte is a Potawatomi scholar-activist at the forefront of Indigenous climate change studies. His work spans the topics of food injustice and food sovereignty, gendered vulnerability and resilience, and more. He has written about how governance models for solar geoengineering often exemplify political obliviousness of Indigenous peoples and how recognition respect for Indigenous sovereignty is crucial.[1] In this interview, we talk more about what free, prior, and informed consent (FPIC) means and whether a decolonized version of geoengineering is possible.

HJB: You've been thinking about Indigenous perspectives on geoengineering for about a decade now—how have your thoughts about it shifted?

KPW: In some of my earlier work, I focused on issues of consent and issues of shared governance. And what I had expected was that there would be more scholars who would take up the different and increasingly complex dimensions of those issues. In a lot of ways, that hasn't happened, at least related to Indigenous peoples. So more recently, I've been trying to bring out some of those more complex dimensions. I've been really trying to bring out some of the deeper issues that are part of this discussion about whether Indigenous peoples embrace or consent to geoengineering or geoengineering research, especially the issue of colonialism. There are certain discussions of colonialism and the history of colonialism that still aren't happening. And so I've been trying to work more to build awareness of the critical aspects of the issue of colonialism.

HJB: I was struck by a piece you wrote about the roles for Indigenous people in Anthropocene dialogues—"roles Indigenous peoples can play *in absentia* or roles they are strongly expected to play when they show up," as you wrote,[2] as disruptors or affirmers, as "*skeletons* in the closets of conservationists who seek to protect islands of the holocene," or as Holocene survivors. You noted that an Indigenous collective hasn't held their own Anthropocene-themed events—and you ask, "In the absence of Indigenous peoples themselves finding value in creating our own dialogues on the anthropocene, what value *for us* does our participation as disruptors, affirmers, skeletons, and holocene survivors—or any other pre-scripted roles—really have?"[3] Solar radiation management (SRM) is also one of these high stakes, megaconcepts, as you call the Anthropocene, that gets proposed without the involvement of Indigenous groups. Do you see those same scripted roles at work with geoengineering? Or is it different from the Anthropocene discourse?

KPW: When it comes to Indigenous involvement in geoengineering discussions, there are a couple of fairly complicated matters to keep in mind.

The first one is that when we refer to Indigenous people, we're referring to some four hundred million people organized into many thousands of societies, nations, and peoples who exercise political and cultural self-determination and who have distinct linguistic, cultural, and intellectual traditions. And so on the one hand, whenever we're discussing a topic like Indigenous people and geoengineering, there are always diverse Indigenous peoples and particular situations that make it hard for there to be such a thing as a unified Indigenous voice.

Yet on the other hand, there's the Indigenous international and global movement that works to advocate for common types of issues and for rights and for other legal protections that would help nearly all Indigenous peoples. And so oftentimes, Indigenous advocates like me talk in a very general sense. While we refer to Indigenous people in this more global sense, we do not seek to betray the inherent diversity, locality, and particularity of Indigenous populations living in different places and experiencing different situations.

A second matter is that in a lot of discussions on geoengineering outside of Indigenous contexts, there was this concern to the effect of "how dare anybody propose the idea that humans play this large role in shaping the environment and shaping the climate system?" This concern often comes across as the idea that this is the first time in which technology had sort of got to the point where people could be asking these types of questions about human collective action and considering these types of solutions to climate change. But for a lot of Indigenous people, the idea of geoengineering—the idea that humans have deliberate regional, continental, and global impacts on ecosystems—is not so new. It's not a shocking idea by virtue of the level of impact being suggested. In fact, you see a lot of Indigenous persons telling their own histories as ones in which their peoples play huge roles in shaping regional ecosystems through their cultural, economic, political, spiritual, and other activities and governance systems.

And so what's been kind of tricky to communicate is that, on the one hand, Indigenous people are not shocked about the idea of geoengineering. On the other hand, there are many Indigenous voices that are speaking out against geoengineering, and I've been trying to convey to people that there's not necessarily an inconsistency here. And so we should be able to say, "Look, geoengineering is something that we're perfectly comfortable with as a concept and an idea and even a potential solution, but there are definitely going to be some Indigenous people that are very against how it is being discussed right now as part of the solution to climate change and climate injustice."

A third matter—and I think this gets closest to what you brought up with respect to my work on the Anthropocene—is that in any form of recent environmentalism, from conservation to geoengineering, you almost always see Indigenous people being used for the purposes of non-Indigenous environmentalists. It's just a common theme—from the well-known nostalgic accounts of Indigenous people in the U.S. environmental movement, such as the "crying Indian" commercial, to more current ones we see in other aspects of environmentalism, such as media portrayals of Indigenous vulnerability to climate change impacts or extractive industries that present romantic accounts of Indigenous victimhood. We're constantly seeing Indigenous people being caricatured, stereotyped, or used instrumentally to serve environmentalist agendas that *often* do not—at the end of the day—seek to support Indigenous agendas, including treaty rights, FPIC, or decolonizing environmental education.

Similarly, in some of the discussions on geoengineering, there's this idea that if it can be shown that there is one Indigenous people or a set of Indigenous peoples who really embrace research on geoengineering, then it must be OK. That is, it must be morally acceptable to engage in geoengineering. Or we get the idea that geoengineering is actually the only thing that might save some Indigenous people, especially those who are facing immediate or near climate change threats that will potentially require them to relocate permanently or that will cause serious problems in their economies or actually stop some of their cultural maintenance and revitalization.

So there's this idea that actually what people who advocate for doing further research on geoengineering are really doing is trying to save those who are most vulnerable. And given in a lot of work I've been part of in the scientific realm, we pretty much know that Indigenous people are among the groups who are facing some of the most immediate—but also the most severe—long-term threats. I want to argue that saving Indigenous peoples is actually not an adequate way to understand or frame Indigenous concerns about geoengineering or Indigenous involvement in geoengineering discussions.

HJB: When I've been in these governance fora—like academic or policy-oriented discussions on geoengineering governance where Indigenous participation is brought up—it's usually in the framework of FPIC. You've talked about a more radical approach beyond being integrated into this approach and why FPIC isn't really adequate. What's a vision for a more radical form of genuine participation?

KPW: Let's back up a bit and look at the history of this. The emergence of geoengineering as a serious solution, or part of the solution, to lessening certain climate change impacts to buy time for serious mitigation to occur came around the same time that the global Indigenous rights movement was really building steam. And so you see within the United Nations then, in 2007, the adoption of the U.N. Declaration of the Rights of Indigenous People (UNDRIP), and one of the key norms in that is FPIC. And so that really reinforced that standard for nation-states as well as for organizations or anybody else that's doing things that could impact Indigenous peoples—that there has to be a process by which FPIC is established.

Now, FPIC is extremely complicated in practice. It can refer to contexts like research studies where a researcher from a non-Indigenous university would then work with an Indigenous people and engage in a thorough process to ensure that the Indigenous people have opportunities to consent or dissent. It could also refer to national policies, like if the United States creates a statute that has something to do with Indigenous peoples, do the Indigenous people have opportunities to consent or dissent from that?

Pertinent to the topic here, FPIC is also taken to refer to larger global issues. Like if a consortium of countries or a large global movement is established to support and finance something like research in geoengineering, were Indigenous people at the table? Did Indigenous peoples have a chance to offer their FPIC to the actions or policies of the consortium or movement? Here FPIC is extremely tricky to understand in terms of its ultimate meaning, especially given the diversity of Indigenous peoples and their different levels of empowerment for being able to voice their concerns and be at the table when dealing with nation-states, corporations, research organizations, and others.

Now, what we're seeing right now are some things that a lot of non-Indigenous scholars working on geoengineering have yet to fully address. On the one hand, there are Indigenous groups, especially Indigenous organizations, such as the Indigenous Environmental Network, that are very adamantly against any form of geoengineering, and they probably say explicitly they dissent. And so I think a lot of the most vocal Indigenous voices and leaders are very much against it. They say they do not consent, they dissent from geoengineering, and then they have a variety of important reasons.

On the other hand, you have these cases like the Haida Gwaii ocean fertilization experiment in which that First Nation in Canada worked with someone in the private industry to engage in ocean fertilization. It was taken up by some people as a test case for Indigenous advocacy of geoengineering. And so the idea was that that First Nation bought into ocean fertilization because it was proposed to them as a solution to help curb the decline in their salmon populations. Of course, salmon is a species traditionally and historically important to the First Nation and part of its current economy. And people will then say, "Oh, well, that's one of these cases where this is a group that consented to something pretty analogous to geoengineering." And I think there are other cases of this as well out there.

Do Indigenous people consent to geoengineering, or do they not? There's a lot of confusion out there. I've seen people use all the different examples I've described to make the case either way. And so what I would argue is that people are misunderstanding a few key points.

The first point is that if you look at the growing literature—whether it's in the Intergovernmental Panel on Climate Change (IPCC) or in the U.S. Global Change Research Program—on what's known about Indigenous people and climate change risks, for Indigenous people, what makes a climate change impact or potential climate change impact risky is not the impact itself. It's not the sea level rise itself; it's not the extreme weather event itself; it's not the drought itself. Rather, it's the fact that for each Indigenous people, they're likely in a situation where the nation-state that they neighbor had drastically reduced their territory and essentially forced them into a smaller area of land to make way for extractive industries and to make way for commercial agriculture and forestry, settler cities and towns, recreational spaces like national parks, and all sorts of other settlement activities. And so the Indigenous land base is small, and most of the same policies and moral norms that permitted colonial domination to occur are ones that continue today to make it harder for quite a few Indigenous peoples to adapt to climate change impacts.

In North America, a lot of Indigenous people claim that the reason they're vulnerable to issues like sea level rise or increases in water temperature is because the United States or Canada still doesn't respect their treaty rights or the United States and Canada still haven't come to terms with the fact that they have forced Indigenous people onto areas of land that are too small for any Indigenous people to be able to adapt. So they're more vulnerable by having to live in small reservations or small villages. And historically, many of the climate change impacts we're concerned about today would not have been a problem for most Indigenous ancestors.

So the argument that geoengineering is actually intended to sort of save or support Indigenous people is actually hard to maintain because it's not temperature rise or unpredictable precipitation that are really the problems. And whether or not those things even happen, Indigenous people are still going to face environmental issues due to the types of things that I just described, including the continued intensification of extraction in a lot of Indigenous territories, right?

So when somebody says, "Do Indigenous people consent or agree to geoengineering?," it's a bit strange because a lot of Indigenous leaders are actually thinking, "Well, wait a minute. We're still trying to decolonize treaty rights. We're still trying to increase our land base. We're still trying to establish respect for our cultures and way of life among neighboring settler populations." And so to suggest that it comes down to an issue of what do Indigenous people think about geoengineering is to ignore all those other factors that foster environmental injustice. This feeds into a key issue that you referenced, which is whether it is even worth the time for Indigenous people to be at the table in these discussions about geoengineering if those discussions are not willing to cover these other concerns that I've raised. For Indigenous peoples, I would argue, those concerns are more to the point regarding environmental and climate justice.

Regarding the case of Haida Gwaii, what I've seen in the journalistic and I think some of the academic or quasiacademic coverage is that people are very intent to understand what are the tribal members' views on ocean fertilization. They somehow think that conversations about geoengineering and its governance will be improved if more is known about what tribal members think about the salmon recovery project. But nobody ever addresses the fact of why the First Nation is in the situation of being concerned about salmon populations in the first place—even though some of the academic coverage I have seen gives ornamental accounts of the history of Canadian colonialism. And it's not climate change necessarily that is behind the problems with salmon populations; it's the continued political situation that the First Nation faces in relation to the Canadian settler state. And so whether or not ocean fertilization continues to happen there, Indigenous persons are still going to be at risk for environmental issues until Canada changes its overall aboriginal legal and policy framework.

HJB: What about the ways in which participation can be burdensome? On one hand, participating in proposed processes of climate engineering governance can be totally inadequate if the participation enters after the technology has already been framed, shaped, or developed—since people aren't being able to develop the ideas at the outside and articulate and frame what they encompass and what their goals are. So more participation in the earliest stages seems crucial. But on the other hand, participation pragmatically can be a burden for people who are on the front lines—not only of climate change but of adverse development, of extraction, of all these things. People are often working with limited capacity to address all the things that are going on. So I'm wondering if there's a tension there where participation in the formation of these concepts is really important—but people are dealing with so much; should they really be taking their time on something speculative, something that can siphon time and energy from their very immediate and material struggles?

KPW: I think it's unreasonable to expect Indigenous people to participate at forums or in discussions where things are already defined in a way that doesn't allow Indigenous people to meaningfully bring the issues that matter to them to the table. I think you will find some Indigenous people who endeavor to be at the table in these discussions—and they have a very heavy burden to bear, given that there's no possible way they can represent everybody. But you're actually not seeing, with respect to geoengineering, a lot of active engagement and participation simply because I think a lot of Indigenous leaders and Indigenous activists see that the discussion is already preframed in such a way that it's not a good use of their time to engage. So non-Indigenous persons shouldn't actually be asking to bring Indigenous persons to the table. Rather, they should be asking how to make the discussions meaningful to Indigenous peoples. This latter question requires a lot more involvement of Indigenous persons in the very planning of what topics should eventually be the ones people take up at the discussion table.

HJB: Gender and particularly race have been pretty much absent from geoengineering scholarship and discussions, and I was wondering if you had any thoughts about how to better address them.

KPW: When I look at the geoengineering discussions, it's entirely male-dominated in terms of many of the scientists. Many of the conceptions of control are classic aspects of patriarchal systems and also a conception of knowledge that doesn't reflect the way in which Indigenous knowledge systems are gender diverse. And so while I haven't talked about this or thought about this too much in relation to geoengineering, I think that gender is actually a huge issue from an Indigenous perspective but also from other perspectives given the types of background issues that I just described. One of the aspects of colonial racism, and I think this is pretty well documented by the evidence that we have, was the undermining of

Indigenous people's gender systems, many of which were not binary gender systems and were in fact systems that embraced gender fluidity and diversity.

When the United States created the boarding schools, for example, they taught Indigenous students heteronormative values. The United States worked to install a kind of patriarchy in tribal communities, hence discriminating against Indigenous women and two-spirit persons. And so you get issues like the prevalence of rape and sexual violence against Indigenous women being much higher than with women in other groups, and a lot of that has to do with problematic U.S. legal policies that basically erase the lives of Indigenous women. Professor Sarah Deer's work has been critical in building awareness of these moral and legal problems.[4] And in Canada, obviously there has been an active movement to address the high prevalence of murdered and missing Indigenous women and girls. Professor Sarah Hunt has important work on this injustice.[5] So you have this context where part of colonialism is an undermining of Indigenous gender systems and then a sexism and a patriarchy that basically causes violence and injustice in many Indigenous communities. And so some of the most important work right now in Indigenous peoples' advocacy and scholarship is linking gender and environmental issues.

HJB: Carbon removal and so-called negative emissions have been in increasing attention lately. I was wondering, given the experiences Indigenous people have with removal, with facilitated extractive industries, large-scale agriculture, and so on, . . . how do you see participation in carbon removal proceeding?

KPW: There's one thing that has particularly bothered me within the overall environmentalist world—including people who are politically conservative but who are making arguments and doing things related to the environment. It's this assumption that people who are more on the left—or who are proposing forest conservation or regulations on extractive industries or carbon removal—are somehow more in line with what Indigenous people want. And oftentimes, I think that assumption just ignores history from at least some of the Indigenous perspectives I am most familiar with. The conservation movement or the national parks movement or the preservation movement—all these movements have actually been just as damaging to Indigenous people as extractive industries. I don't actually think that's a very controversial point; I think that bears out in most historical studies. For Indigenous people whose economies today are based in extractive industries, this dependence was actually a product of an earlier point in the twentieth century when countries like the United States actually reengineered those tribal governments to be more tied to extractive industries and commercial agriculture as economic means. The works of scholars like Professor Melanie Yazzie and Professor Dana Powell are currently addressing some of these histories.[6]

And so when we look at these different histories, right, we realize that actually when somebody proposes a solution to climate change or proposes a certain way of engaging in something like geoengineering, Indigenous people immediately might actually be rather ambivalent about whether that could be good or bad for their particular community or nation. And so even today with programs like the REDD Program, we see that it's in some cases repeated problems of displacement and disrespect for Indigenous land tenure that are pretty much no different from what conservationists did in the nineteenth century. Scholars like Professor Betsy Beymer-Farris, for example, have been showing this in their work.[7]

There's kind of this sense of a lack of appreciation for history when people propose solutions, and this gets into this issue of temporality where if you look at a lot of the concerns that Indigenous people have with legal and policy frameworks that govern climate change, many of these issues will not be solved in the short term. So Indigenous people are huge environmental advocates. At the same time, many of the solutions we seek are going to take a long time to put into practice. And so we're also in this for the long game, even though our communities are oftentimes hit very hard by a number of environmental issues, from nutritional issues due to declines in fish and plants and animals or to related psychological issues such as historic trauma and depression.

And so we're facing these issues head-on and engaged in a number of short-term actions to palliate as much as possible these extreme issues that we're facing. At the same time, we're heavily invested in long-term legal and policy change. And so when it comes to something like, say, carbon removal or another one of these climate change / geoengineering ideas, I think for a lot of Indigenous people, the question is, "Will they be implemented in legal, policy, and other frameworks that make it possible for tribes to be able to provide voice, to participate in, to say no to, or to embrace in ways that don't force tribes into the same types of trade-offs that we had to face going back a long time ago with the conservation movement or with the reengineering of tribal governments for extraction?" So are these actually going to be solutions that we can engage in like everybody else, or do we have to be put in a funny situation?

A case of this came up fairly recently. I won't say when and where because it's a bit of a caricature, but it just gives you an example. In one particular area, some activists and policy makers were in the process of creating and passing a policy on carbon costing. The policy that they want is very much, I think, in line with what tribes would want to get on board with. When they were a month away from whatever political referendum or decision was going to be made on this, the activists wanted to speak to an audience at an event that would have some tribal leadership at it. Yet the activists started off by saying, "The window is closing very rapidly for tribes to be involved in this, and so we hope that this is something that you can immediately get on board with." When somebody says that to you, it portrays that there's just not enough time for Indigenous leadership to actually figure out what's going on and to build trust and to build an enduring relationship. It triggers the painful reality that settler and Indigenous groups are still not connected enough so that it would have been possible for Indigenous persons to have been involved when this policy on carbon costing was developed in the first place.

In some of my work that I've done interviewing tribal people, they actually argued that what consent means or what prior participation means is at the conception of the potential collaboration—right when you start working together. And so with a lot of these potential solutions, it could be that there are many tribes that would embrace it, and there could be people who are going to be against it—but the question is whether those solutions are going to come about in a way that doesn't naively or blatantly repeat the mistakes of the past. Indigenous persons are of course going to be ambivalent in endorsing solutions that are superficially attractive, whose design and implementation strategies repeat the mistakes of the past.

HJB: If you look at what some of the projections for carbon removal are, they imply that carbon will be captured—from power plants, ethanol facilities, and so on—and sequestered in great amounts under the North American prairies, among other places. For example, they imply a whole network of pipelines for carbon dioxide removal (CDR) and storage in both abandoned oil wells and saline aquifers.[8] I was looking at these maps and thinking about what you've written about "terraforming of Indigenous territories that renders contemporary Indigenous features of landscapes invisible"[9]—I mean, this would be yet another configuration of domination of these Indigenous lands, though potentially one that could reduce some forms of climate risks while creating others. What do you make of this imaginary of putting all this captured carbon back underground?

KPW: The development you describe is—in possibly unfortunate ways—an example of what I am concerned with. In the different but analogous case of the Dakota Access Pipeline, the builders, of course, saw the landscape as primarily a resource in the sense that it could be terraformed to make way for the pipeline. Their strategy for minimizing public, Indigenous, and various other types of involvement was to build most of the pipeline on private property, save for only those areas, like crossing the Missouri River, where permits would be involved. According to the builders, they did take seriously concerns about Indigenous cultural heritage and Indigenous consultation. Regardless, the builders and their political supporters were miffed by the idea that the Standing Rock Tribe and other Indigenous peoples had concerns about the pipeline given construction occurred "off reservation." There was literally no understanding whatsoever of the total landscape that matters to the tribe (which extends far beyond "the reservation") and the years of resistance the tribe has engaged in to protect its members' self-determination and ways of life.

New books that came out in 2019 by Professor Nick Estes and Professor Dina Gilio-Whitaker both show the long histories of Indigenous resistance that one must understand to understand the struggle against the Dakota Access Pipeline.[10] In terms of carbon removal, if proponents of it already approach landscapes as "prairies" or take for granted how power plants got to be in the places they are in the first place, then I don't think it will be possible to implement carbon removal schemes that are beneficial to Indigenous peoples. When people grasp Indigenous peoples merely as "groups to be consulted" or liabilities or stakeholders in terms of historic heritage, I do not think that they can possibly have productive or reconciliatory relationships. I can imagine many scenarios where pipelines and other infrastructure that support carbon removal further displace and disempower Indigenous peoples, degrade Indigenous homelands, and disrespect Indigenous cultures and histories.

HJB: Audra Mitchell has an excellent piece on decolonizing the Anthropocene, where she writes that it's important to resist "solutions" to the Anthropocene that involve colonial logics, like geoengineering.[11] Mitchell writes that "it stands to reason that if colonial violence is a major driver of the Anthropocene and its more deadly effects, it is foolhardy to reproduce this logic." But then she says that "the implication is not that such projects should be ruled out, but rather that careful reflection is needed to ensure that they do not reproduce colonial violence, but rather focus on alternative forms of (in)habitation."[12] I'm curious if you have anything to add to Mitchell's point. Is it possible to imagine a decolonized version of geoengineering?

KPW: First, I would encourage anybody to read Professor Mitchell's piece and overall work. In the work most relevant to mine, Mitchell does a fantastic job bringing together diverse Indigenous intellectual traditions with many others to address the topics of human-nonhuman relations, extinction, and climate justice. Mitchell is also a member of the Creatures Collective, a group of scholars who do very creative collaborative work together and which includes scholars such as Professor Zoe Todd who has done significant work on colonialism and the Anthropocene.[13]

Geoengineering is obviously a topic that bears heavily on how we imagine the future. If you look at a lot of writing and scholarship on geoengineering, it has this assumption about how we imagine the future that you see in other aspects of climate change advocacy and climate justice. The assumption is the idea that if we as humans don't do anything about climate change, there's going to be this future disaster, dystopia, or dreaded period. Humans certainly don't want to see this happen. And so you get a lot of people writing about how if humans don't do something now, then we're all going to see the extinction of the polar bear, we're all going to see the loss of coral reefs, or we're going to see a number of types of novel ecosystems replacing ones that had been around for a certain period of time and that humans, for whatever reason, see value in preserving. This is not to mention the kind of catastrophic impacts on cities and rural communities and many other places of intense or extreme weather and ecological shifts.

How might this assumption about climate change strike a number of different Indigenous peoples? For my tribe or for many others, we're already on the other end of that—actually. The current situation we're in is actually like a dystopia, in a sense. If you had asked our ancestors, if you'd have told them what today's times are like—where many Indigenous people are highly disempowered, we've lost relationships with hundreds, if not thousands, of plants and animals, and the majority of habitats our ancestors would have recognized on Turtle Island are completely eliminated and have been terraformed by the United States to reflect US cultures, US economies, and US values. Today, it's kind of like a science fiction film for many of us, where we awaken in this environment where many of our peoples are boxed into reservations or heavily limited in their capacity to empower themselves in urban areas. Almost every law or policy works against us, overtly and subtly. In countries like the United States, the majority of people actually don't even know that there are tribes that are recognized as sovereign by their own country, the United States. The educational system largely brainwashes people into ignorance about Indigenous peoples' own histories and settler violence against Indigenous peoples. Professor Candis Callison has shown that Arctic Indigenous leaders had to push many environmentalists to realize that there are people who have lived in the Arctic for centuries, since some environmentalists kept focusing on the plight of the polar bear in isolation from the actual peoples of the arctic.[14] And so we're in this highly dystopian environment, at least from our ancestors' perspectives.

And so what we're trying to do is not necessarily avoid a dystopia, because we're already in it. We're trying to figure out how to best empower ourselves in this highly impossible situation that we continue to survive through. Many Indigenous people are actually able, through hard work and creativity, to continue to somewhat flourish under these circumstances.

What I think is ironic about a lot of geoengineering discourse is that the current situation that geoengineering is trying to preserve or save is actually a dystopia for some people. Members of dominant populations are trying to avoid their dystopia by preserving our dystopia. And that oftentimes is a very uninviting way to engage in a conversation because, for Indigenous people, we often want to talk about the fact that we're still living in this situation where it's almost impossible for us to consent or dissent from anything. The legal policy barriers, the continued discrimination, and the massive habitat changes that have already occurred thanks to colonialism, capitalism, and industrialization—that's actually our starting point. Our starting point is not how do we maintain the current situation but actually, how do we get out of it? For people who are saying that we need to preserve the current situation, that just strikes me as highly uninviting.

Is there a decolonial kind of geoengineering? I think absolutely, but I think it would require us to really redefine what we mean by geoengineering and would require some solutions that I don't really see as ones that could practically be implemented in the short term.

One practical beginning, in terms of the actual discussions that people are having about geoengineering, is to actually couple the geoengineering discussion more deliberately with the issues that are most important to Indigenous people. And in this context, it's not just consent in the FPIC sense. It's related to some of the other things that I brought up having to do with treaty rights, having to do with agreements with states and corporations, having to do with issues of land tenure, having to do with issues of being able to live in ancestral territories that were dispossessed from Indigenous people but are nonetheless economically and culturally significant. Of course, all these examples are about consent too—but in a much deeper sense than I have seen invoked in geoengineering discourse. I think a discussion where geoengineering is embedded in that context is one where people like, say, those in Haida Gwaii would be much more empowered to make meaningful change. The current conversation about geoengineering seems very much about, depending on what one's goals are, that one either wants there to be Indigenous people who say, "Yes, I want geoengineering" or "No, I don't." In geoengineering discussions I have been part of, people are sort of cherry-picking different case examples of Indigenous peoples to support whether they advocate for or are against geoengineering.

Another way of thinking about the decolonization of geoengineering is to go into what a lot of tribes are doing, which is to try to reinvigorate their longstanding intellectual and governance traditions of organizing their governmental systems around the seasons and around lunar and other types of cycles. In these traditions, the very idea of governance is trying to approximate the dynamics of ecosystems instead of creating a form of governance that satisfies humans and then trying to figure out how to make it consistent with the dynamics of ecosystems that it wasn't designed to be compatible with in the first place.

And so we're seeing a lot of Indigenous people really try to suggest that issues like treaty rights or political reconciliation need to be based on understanding governance as a matter of redesigning the actual institutions so that they're adaptive to and responsible in relation to the dynamics of ecosystems. And in a sense, you can look at that as a form of geoengineering because it would be changing the entire model of governance and how humans, through institutions, impact the environment. This would change how we think of things like carbon footprint or energy or other topics. And so I think for some people, if asked about a decolonized form of geoengineering, they could go in that direction.

Notes

1 Kyle Powys Whyte, "Way beyond the Lifeboat: An Indigenous Allegory of Climate Justice," in *Climate Futures: Reimagining Global Climate Justice*, ed. Kum-Kum Bhavnani et al. (Oakland: University of California Press, 2019); Kyle Powys Whyte, "Now This! Indigenous Sovereignty, Political Obliviousness and Governance Models for Solar Radiation Management Research," *Ethics, Policy & Environment* 15, no. 2 (2012): 172–187; Kyle Powys Whyte, "Indigenous Peoples, Solar Radiation Management, and Consent," in *Climate Justice and Geoengineering: Ethics and Policy in the Atmospheric Anthropocene*, ed. Christopher J. Preston (Lanham, Md.: Rowman & Littlefield, 2016), 65–76.

2 Kyle Powys Whyte, "The Roles for Indigenous Peoples in Anthropocene Dialogues: Some Critical Notes and a Question," *Inhabiting the Anthropocene*, January 25, 2017, http://inhabitingtheanthropocene.com/2017/01/25/the-roles-for-Indigenous-peoples-in-anthropocene-dialogues-some-critical-notes-and-a-question/.

3 Whyte, "Roles for Indigenous Peoples."

4 Sarah Deer, *The Beginning and End of Rape: Confronting Sexual Violence in Native America* (Minneapolis: University of Minnesota Press, 2015).

5 Sarah Elizabeth Hunt, "Witnessing the Colonialscape: Lighting the Intimate Fires of Indigenous Legal Pluralism" (PhD diss., Department of Geography, Simon Fraser University, 2014).

6 Dana E. Powell, *Landscapes of Power: Politics of Energy in the Navajo Nation* (Durham, N.C.: Duke University Press, 2018).

7 Betsy A. Beymer-Farris and Thomas J. Bassett, "The REDD Menace: Resurgent Protectionism in Tanzania's Mangrove Forests," *Global Environmental Change* 22, no. 2 (2012): 332–341.

8 Daniel L. Sanchez et al., "Near-Term Deployment of Carbon Capture and Sequestration from Biorefineries in the United States," *Proceedings of the National Academy of Science* 115, no. 19 (May 2018): 4875–4880.

9 Kyle Powys Whyte, "Indigenous Science (Fiction) for the Anthropocene: Ancestral Dystopias and Fantasies of Climate Change Crises," *Environment and Planning E: Nature and Space* 1, nos. 1–2 (2018): 224–242.

10 Nick Estes, *Our History Is the Future: Standing Rock versus the Dakota Access Pipeline, and the Long Tradition of Indigenous Resistance* (London: Verso, 2019); Dina Gilio-Whitaker, *As Long as Grass Grows: The Indigenous Fight for Environmental Justice from Colonization to Standing Rock* (Boston: Beacon, 2019), 224.

11 Audra Mitchell, "Decolonising the Anthropocene," Worldly, March 17, 2015, http://worldlyir.wordpress.com/2015/03/17/decolonising-the-anthropocene/.

12 Mitchell.
13 Mitchell; Audra Mitchell, "Revitalizing Laws, (Re)-Making Treaties, Dismantling Violence: Indigenous Resurgence against 'the Sixth Mass Extinction,'" *Social & Cultural Geography* (2018): 1–16; Zoe Todd, "Indigenizing the Anthropocene," in *Art in the Anthropocene: Encounters among Aesthetics, Politics, Environments and Epistemologies*, ed. Heather Davis and Etienne Turpin (London: Open Humanities, 2015), 214–254.
14 Candis Callison, *How Climate Change Comes to Matter: The Communal Life of Facts* (Durham, N.C.: Duke University Press, 2014).

6

Recognizing the Injustice in Geoengineering

—————————————————————•

Negotiating a Path to Restorative Climate Justice through a Political Account of Justice as Recognition

DUNCAN McLAREN

Climate change is not simply an environmental problem susceptible to economic or technical "solution" but a cultural phenomenon that structures contemporary life.[1] If we are to suggest appropriate responses, we must take account not only of the ways it conditions politics, economics, and culture but also of its political, economic, and cultural origins. In this chapter, I begin with an understanding of climate change as a material product both of the cultural and economic transition to modernity via colonialism, slavery, and industrialization and of particular formations and contestations of politics and power constructed on fossil fuels.[2] These histories converge in the dominant social imaginary of industrial, administrative (neo)liberalism[3] that constructs climate change as a technical problem susceptible to market-based and technological solutions. It is accepted that the distributed effects of climate change raise questions of justice but only within the liberal distributive paradigm. Responses such as carbon trading and technological measures such as carbon capture and storage (CCS) dominate the narratives of both climate scientists and politicians (and most other stakeholders), while even restraining the growth of emissions has proved painfully slow.

As a result, geoengineering responses are coming to the fore in contemporary debate over climate change. Large-scale carbon geoengineering in the form of bioenergy with carbon capture and storage (BECCS) is foreseen in most modeling of safe climate pathways,[4] and there is growing public and media debate over the case for solar geoengineering's ability to directly reduce ambient temperatures.[5] Moreover, solar geoengineering in particular is presented as a means of supporting climate justice by disproportionately reducing the climate risk faced by the poor and vulnerable.[6] While other commentators have challenged the paternalism implicit in such claims and highlighted the importance of improved participation by disadvantaged groups, to enable procedural justice,[7] the debate largely remains within the liberal paradigm and the dominant social imaginary.

Instead, I argue, a critical reading of these histories and the conditions of climate change from the perspective of cultural political economy (CPE)[8] exposes a coevolution of the dominant imaginary with climate policy and geoengineering techniques that tends to conceal the extent of climate injustice.[9] Instead of geoengineering solutions, I argue that the history and impacts of climate change suggest a demand for restorative justice on two dimensions: in the relationships between humans and the planet and in the relations between the beneficiaries and victims of colonialism and industrialization. Yet for restorative justice to become a political possibility, a prior step of recognition is necessary. The victims of injustice must obtain recognition as full moral equals and, moreover, must obtain political voice and agency. The very idea of justice as recognition is, however, marginalized in the dominant social imaginary and typically understood merely as a step of moral or political inclusion in the existing system. It can therefore be criticized by radical scholars who portray the pursuit of inclusion, as opposed to transformation of the system, as a betrayal of class or group interests.[10]

In this chapter, I argue that geoengineering techniques are being coconstructed with political regimes inside the dominant liberal administrative social imaginary,[11] acting therefore as sustaining innovations for the political and cultural maintenance of distributed privilege. I seek to illustrate ways in which current geoengineering research and advocacy fail to properly recognize all those that would be affected by it and, moreover, privileges (among other things and typically without question) certain forms of knowledge, expertise, moral theory, and subjectivity. I argue that with a political account of justice as recognition that centers the quest for justice as the primary task of politics, we can establish a foundation that enables a critical political analysis of geoengineering research and advocacy.

The chapter proceeds by briefly describing the emerging coproduction of geoengineering technologies and the ways in which this process misrecognizes groups, epistemologies, and interests. It then outlines a political account of recognition and concludes with suggestions for how our responses to climate change might become more just as a result.

Geoengineering and Climate Change

Proposals for geoengineering are not new. They have been featured in discourses about climate change since at least the 1960s.[12] However, the technologies involved have varied widely, and the forms of geoengineering technology currently under most consideration (BECCS and stratospheric aerosol injection [SAI]) have, in important ways, been constructed within climate policy discourses and in particular within climate modeling practice using earth system models and integrated assessment models (IAMs).[13] More broadly, the deployment of climate models to inform policy development echoes at least three significant features of the dominant modern social imaginary documented by Christopher Groves.[14] First, it is part of an administrative, postpolitical turn to policy formation that is heavily reliant on expert advice and the calculation of budgets and probabilities. Second, it is a managerial approach that seeks to domesticate uncertainty about the future, reducing it to certain parameters and evaluating it strictly in terms of current interests and values. Third, at least within the IAMs, it embodies a utilitarian, market-based approach to choice. The models are designed to deploy technologies to minimize financial and social costs, assuming a high fungibility between different costs and benefits and between different technologies (regardless of divergent side effects, cobenefits, or distributional implications) and more generally, largely ignoring implications that cannot be monetized.[15]

BECCS was introduced into climate modeling around 2006 in models that suggested it would reduce the cost of achieving the desired outcomes.[16] It did so by enabling a more gradual reduction in near-term emissions, and thus it continued exploitation of and profit from fossil fuel extraction and use. While it remains impossible to attribute causation, this appears to have at least contributed to, if not stimulated, the slow progress on global emissions reductions in the past decade. Now models mainly suggest that the desired climate outcomes cannot be achieved without negative emissions (still typically represented as large deployments of BECCS later in the century). Modelers exploring the potential for SAI also suggest that it could reduce the financial burden involved in achieving desired climate targets, bring more stringent climate targets into practical reach, and allow for a less disruptive transition to a low-carbon economy.[17]

The technologies remain largely imaginary. A handful of BECCS pilot plants exist but none of the nature or scale foreseen in the models, and SAI remains an entirely theoretical construct at present. The design of the technologies is therefore a product of the imaginaries that surround them: those of administrative, managerial modeling within a dominant liberal social imaginary. This extends to the conceptions of justice against which geoengineering ideas are assessed (insofar as justice is acknowledged as an important criterion). Utilitarian and consequentialist assumptions dominate. High levels of BECCS might divert land from food production, but the benefits—such as fewer restrictions on air travel as a result of substituting negative emissions for controls on consumption of jet fuel—are implicitly deemed more valuable. Similarly, SAI might result in unpredictable disruptions to local climates in many parts of the Global South, yet the global scale benefits of lower average temperatures are implicitly valued more highly.[18] Most researchers encourage more public education and consultation, but the model of procedural justice is also one rooted in a liberal administrative social imaginary. As long as affected people have had a chance to raise their objections and decision-makers have reasonably considered those representations, minorities may be expected to bear severe costs in return for overall social benefit (such as local communities hosting nuclear waste facilities or Indigenous groups "accepting" oil sand developments).[19] This understanding of justice holds especially where those local communities appear to accept such facilities, however clear it is that such "acceptance" is only because they are otherwise heavily disadvantaged.[20]

In other words, arguments for geoengineering research and deployment rely on the privileging of particular goals (reducing climate risk by achieving particular carbon budgets / temperature change outcomes) and of particular paradigms of knowledge and evaluation (utilitarian, administrative, liberal), which thus puts particular groups and interests in global society above others. The results could be especially unjust if expectations of geoengineering—whether through large-scale negative emissions or through solar radiation management—were to substitute for accelerated mitigation by high-emitting groups and countries and adaptation financed by the rich world. I now turn to the argument for a political account of justice as recognition, interpreting these processes as misrecognition or failure of recognition.

Recognition and Justice

I do not seek to argue that justice is the only substance of politics. For the purposes of this chapter, it is enough to assert that justice is intensely political. Whether in terms of the distribution of resources within and between societies or the inclusion or exclusion of different groups in societies, it is impossible to discuss justice without raising political concerns.

The normative ideal that justice demands that each of us is recognized as a full moral subject in society—by other subjects and by social institutions—is present in much mainstream moral philosophy. In some accounts, it is explicit,[21] and in other accounts, it is implicit. While some scholars understand recognition entirely or primarily as a psychological and individual phenomenon central to the development of personhood and self-respect, others emphasize intersubjective elements and the importance of recognition to caring relationships and love and see the quest for recognition as a source of social movements for inclusion.[22] Efforts have also been made to develop political accounts of recognition, most notably by Charles Taylor.[23] Here recognition is central not only to personhood but to the functioning of multicultural societies. Nancy Fraser[24] has developed recognition as a complement to redistribution as a foundation for emancipatory justice understood as full participation in society (which might be hindered by lack of material resources, cultural discrimination, or both). More recently, recognition has been introduced into environmental justice scholarship as one of the pillars of David Schlosberg's trivalent approach (alongside distributive and procedural justice). Schlosberg[25] is concerned that the victims of environmental injustices are not fairly recognized as either equal citizens or specific individuals and groups with culturally and materially situated needs.

Here I seek to develop this approach further, arguing that recognition offers not only a political but a potentially radical and transformative approach to justice. I argue that this is because recognition simultaneously (1) moves away from the liberal subjectivity of autonomous individuals, treating identity as a continuing, intersubjective, and dialogic process, and (2) acknowledges and welcomes the diversity and significance of the multiple, intersecting dimensions of identity and culture embodied in persons, subject only to a constraint of reciprocal solidarity as a criterion for legitimate claims of difference. In other words, a political account of justice as recognition allows for two critical steps: First, to move outside the dominant liberal social imaginary and accept the likelihood that our current conceptions of justice are too narrowly defined and undesirably culturally specific. And second, to work toward the transformation of society to enable ever-greater political and cultural inclusion, not just for unrecognized humans but for other actors and agents in the world.

Within the liberal paradigm, recognition would demand that we acknowledge the various groups—such as Indigenous peoples, women, or future generations—that are vulnerable to climate and geoengineering impacts and the ways in which institutions and systems may impede their full participation through misrecognition. But a critical political understanding of justice as recognition must also engage with the procedural and substantive terms on which inclusion is achieved. Recognition (following Schlosberg) advocates treatment of all persons[26] as moral equals, recognized and valued *for* their differences, not *despite* them. Demands for recognition of difference, such as those in the Black Lives Matter campaigns, are not primarily a call for equality and inclusion in the existing social and political order, but they are campaigns for reconfiguration of that order in ways that recognize and respect cultural diversity, difference, and history. Like recognition demands from Indigenous communities, they are calling not for integration but for respect.[27] Nor are such demands necessarily rooted in essentialized ideals of cultural identity but rather involve the rejection of the identities imposed by structural and cultural oppression and discrimination and thus constitute demands for the opportunity and agency for groups and collectivities to reconstruct and reconfigure their own identities. Recognition asks social institutions and other individuals to acknowledge—and value or reject as appropriate—diverse aspects of individual or group identities where they are significant to those people as a part of their self-identity, confidence, or agency or as a source of economic, cultural, or other disadvantage. Thus, for example, to respond to racial oppression with a stance of "color-blindness" is as much of a failure of recognition as is active discrimination on racial grounds. To fail to acknowledge differential vulnerability to climate change, or to geoengineering, is a failure of recognition too.

A political account of justice as recognition makes the claim that justice requires both intersubjective and institutional recognition of all persons (and perhaps other agents) as fully equal moral agents or subjects (understood in their own terms and identities) conditioned on a reciprocal solidarity (a mutual recognition of interdependence). The outcome of full recognition is that agency is enabled. The recognized person can fully participate in society. However, such participation is not limited to paternalist inclusion in the existing social order but includes the potential (arguably necessity) of social transformation through the inclusion of the hitherto unheard or uncounted subjects.[28]

Criticisms of Recognition

As an approach to justice and a political philosophy, recognition has faced criticisms from multiple directions.[29] For some, it is too subjective, lacking criteria to differentiate legitimate claims for respect based in identity from unreasonable ones (based, for example, in the identity of white supremacists). For others, it relies on essentializing identity unrealistically (as an unchanging authentic kernel of selfhood), risks reifying identity to the extent that group uniformity is valued over individual diversity, or ignores the way in which subservient identities can be imposed through the exercise of power or habitus. Space precludes exploring and rebutting all these critiques, but here I wish to engage with the view that recognition is not emancipatory but affirmative of the power relations of society. If this were true, recognition would not belong in a critical political toolbox.

Some radical discourses of justice have treated recognition dismissively, as the basis of forms of identity politics that are alleged to weaken and undermine collective political action.[30] Moreover, identity-based claims are criticized as themselves a product of oppression, a sort of false consciousness.[31] I argue, however, that we need to approach such analysis not simply at a constructivist or cultural register but from a CPE register[32] in which the relationships between power and identities and between power and framings can be exposed and examined. With such analysis, we can begin to understand how particular perspectives and framings are granted voices within existing power relations and others remain unrecognized and unheard. A CPE account of recognition also problematizes the processes by which identities are constructed and reconstructed. This is particularly valuable in the face of emerging understandings of human-nature relations in discourses of the Anthropocene: a new geological era marked by human impact. In such discourses, humankind is typically portrayed as a uniform entity, whether powerful (world-making) or vulnerable (to earth system change), rather than as diverse and divided. For instance, advocates of a "good" Anthropocene see geoengineering as one of the tools of a powerful humanity wielded in the interests of humanity as a whole.[33] Others portray humanity as uniformly vulnerable to climatic change: "all in it together," and climate geoengineering as a hubristic boast with unpredictable consequences. While a single humanity in which all are fully recognized and included may be a desirable aspiration, it is far from the current situation.

Moreover, within the liberal paradigm of justice and the administrative social imaginary, ideas of recognition are often either presented as if participation in society on the existing terms defined by those already included would be adequate[34] or criticized as if they could only be interpreted as suggesting such a limited form of participation.[35] Recognition is typically portrayed in thin ways as a partner to rights and procedural justice, implying cultural assimilation of the newly recognized, rather than a means by which such procedures might be enriched or reconfigured in ways that offer intercultural respect.

Critics of recognition too tend to interpret recognition within narrow liberal paradigms and the administrative imaginary, which arguably makes the task of escaping those imaginaries much harder, undermines the possibility of recognition of the alternative values and knowledge systems implied by those critiques, and misrecognizes the concepts of recognition themselves. The versions of recognition that are critiqued in the work of Lois McNay[36] and of Irina Velicu and Maria Kaika,[37] for example, are in some ways extremely distorted, paternalist models in which recognition is entirely controlled and rationed by existing elites or power structures. In the body of this chapter, I seek to develop a richer political account of recognition set in a CPE perspective and illustrate how ideas of geoengineering can reinforce misrecognition and exclusion, leading to the exacerbation of climate injustice.

Developing the Political Account of Recognition

The crux of a political account of recognition is the constitution of citizenship and subjectivity inherent in the concept. Recognition is critical both to self-realization of the human subject and to our full participation in society. The harm of nonrecognition or of misrecognition is both the social status injury arising from exclusion from society and also the psychological harm arising to the personal identity from such rejection or exclusion.[38] A secure identity depends on care, respect, and self-esteem.[39] But identity formation is an *ongoing* intersubjective dialogue, not simply a one-off process for each individual. To support that process requires not only liberal rights but also communitarian relations of care that help us actively recognize and resist the power of structures and discourses to internalize oppression and discrimination in identities, behaviors, and beliefs. Full recognition requires measures both to ensure the status, resources, and voice necessary for citizens to be able to enjoy parity of participation[40] and to support the care, respect, and esteem that enables us, mutually, to build capacities to sustain, repair, or reconfigure our identities in ways that reflect the goals of justice as recognition.

Recognition and nonrecognition do not constitute a simple binary state. Where members of a certain group are completely excluded from the moral community, that is *nonrecognition*. Nonrecognition constitutes its victims as invisible, unheard, and even as "nonpersons." It legitimates instrumental treatment of members of that group (e.g., of animals; previously of Indigenous peoples and slaves; and today in some places of migrants or refugees). Partial recognition can be normatively preferable (and for actors unable to demand recognition, such as future people or nonhumans, partial or "granted" recognition might be all that is possible). Yet partial or inaccurate "misrecognition" is often harmful. Colonial metropolitan science constructed a *homo primitivus* of Indigenous populations, a nonrecognition that legitimated their decimation and the usurping of their lands and resources.[41] Today's scientists and economists misrecognize non-Western peoples rather as *homo economicus*, but the result is still a failure to acknowledge the significance of difference, which leaves those groups doubly disadvantaged by the processes of underdevelopment. More generally, misrecognition persists where groups are included in the moral community of humans merely as "humans" (an extremely "thin" form of recognition), yet specifics of their culture or identities are ignored, such as treating trans people as if they were cisgendered. Another example of such misrecognition is found in geoengineering modeling work that treats those whose cultural identities depend on climatic and environmental features of their territories in the same way as it treats modern Western societies—as if all that mattered in the face of geoengineering were material impacts of climate on food production or bodily security.[42]

Furthermore, mainstream ways of thinking about future people constitute the sort of institutional or structural misrecognition that would be rightly condemned if those concerned were women or people of color. For example, institutional use of economic discounting in considering public investments systematically devalues the costs and benefits accruing to future people in relation to those arising in the present, thus incentivizing a transfer of benefits from the future to the present and of costs from present to future.

The intergenerational and international challenges are similar in that recognition here cannot rely on direct intersubjectivity. Yet direct intersubjectivity is not essential. Recognition might be extended through an abstract or generalized form of recognition of the other established through a cognitive imagination of empathy for the other,[43] resembling a globalized form of care relations that gives "attention to actual differences between persons and groups" and "resistance to universalizing all into an abstraction of the ahistorical, rational-individual-as-such."[44] Similarly, in contractualist terms,[45] mutual recognition requires that we give others the kind of deliberative consideration of their relevant interests to which they are entitled—where that entitlement is a general principle that no one can reasonably reject.[46] Such expectations mean that descendants of slaves can legitimately feel resentment and demand acknowledgment of the harms done to them and restorative apologies or compensation from those complicit in inheriting the benefits of slavery. The same would appear to hold for future people experiencing the effects of climate change or geoengineering.

Arguably, some abstraction moreover helps ensure that we do not, through recognition, essentialize unjust forms of identity produced by oppressive habitus.[47] And within a political approach to recognition, we might also draw on more communitarian concepts of intergenerational commitments and responsibilities rooted in group identities and norms. In this understanding, our care for future people is based on a desire to see collective projects, cultural traditions, and norms preserved or maintained.[48]

So recognition can extend to people remote from us in space and time. Where the separation is greater, the process is more cognitive and less dialogic and recognition is granted more than it is demanded. Yet it still offers potential for emancipation of previously oppressed groups. Moreover, it introduces new voices and interests into the politics of climate change. This could be transformative. If we understand politics as a process through which a community establishes or changes norms, rules, and institutions, then politics is about freedom, or agency, within a diverse common realm, and recognition enables the previously unheard to participate as equals in public affairs.[49] Not only does such participation offer enhanced justice; it also stimulates the reconfiguration of norms and institutions and potentially challenges the dominant social imaginary.

A political concept of recognition as justice clearly has to mean more than a paternalist granting of recognition by the dominant group, admitting a new minority into the moral or political community as long as they adopt the values of the majority. That would be misrecognition. It ideally therefore implies the demanding of recognition (by the unrecognized group and/or their representatives) through agonistic resistance, conflict or struggle, and critical, subsequent dialogue between the groups so as to seek to understand and respect each other's values. As James Ingram[50] argues, this constitutes politics as a process of never-ending enhanced inclusion. But such dialogue must be rooted in a mutual recognition of the values of inclusion, solidarity, and "living together"; otherwise, it has no common foundation or even reason for taking place. Rather than demanding a broad commitment to liberal values, rights, and duties as the foundation for moral inclusion, this analysis rather suggests a single universal foundation of a commitment to mutual recognition or solidarity as the irreducible minimum, on which dialogic processes can be built.

This account does not deny that (mis)recognition can be deployed—intentionally or emergently—as an exercise of power, but it understands recognition also as a tool to expose and resist unjust power structures.[51] The agency offered by recognition is not simplistic or essentialized, but one still conditioned by habitus, discipline, and elite power (all of which may result in forms of misrecognition). Nonetheless, recognition decreases the asymmetry of power and empowers different cultures and different ways of knowing, contributing to epistemic justice.[52] And above all, it increases the potential for the newly empowered agents to generate disruptive dissensus,[53] challenging the terms of the social contract, reconfiguring the moral community, and transforming the dominant social imaginary.

This account of recognition consciously echoes Jacques Rancière's understanding of postpolitics and politics proper.[54] Yet Rancière tells us little about the moral obligations of citizens already within the community, and his model cannot speak to circumstances in which the unheard have no practical means to express their demands (e.g., future people or nonhumans). So although postpolitical framings dominate the issues of climate and climate geoengineering,[55] to overturn these requires granted forms of recognition for future people and possibly even also for nonhumans and environmental systems. By establishing recognition as a relational process that triggers caring attachments, this account suggests richer forms of political subjectivity. By problematizing and politicizing identity, as recognition encourages us to do, we end up with a possibility of pursuing justice that otherwise seems impossible in individualized and essentialized accounts of identity or their critiques. In contested processes of recognition, we can engage with both the multiple dimensions, values, and relational and situated factors that are constitutive of our sense of self and agency and the influences that structure and condition them. Identity projects and associated recognition can be, politically, tools of governmentality and discipline or tools of revolution and transformation rather than necessarily one or the other. But without engaging in them and with the cultural, political, and economic regimes that condition them, the possibility of justice is denied to all but those already recognized within the dominant social imaginary. Technology also plays a critical role in the maintenance of the dominant social imaginary. Technological progress is understood as a source of economic growth and the means by which social or environmental crises can be "fixed" (or more normally, displaced). Geoengineering technologies are typically imagined in ways that reinforce such narratives of progress, especially in ecomodernist discourses that portray them as tools of planetary governance in the Anthropocene.[56]

The psychological dimension of mutual recognition is also critical to this political account. It places moral and psychological vulnerability into the center of the human condition. We are vulnerable to (and dependent upon) our fellow humans, as only they can grant the recognition we crave. The vulnerability of our personal worlds and our attachments require us to face up to our condition of dependency on the care of others.[57] When vulnerability itself is recognized, it underpins interdependency and care as a political challenge. A care-based imaginary rooted in a political approach to recognition that responds also to activist and movement demands for recognition[58] would acknowledge vulnerability and interdependence and stimulate a moral obligation to recognize both our complicity and our capacity to act to make recompense.

In this respect, demands for recognition are rooted in the protection or restoration of our constitutive relationships of care, attachment, and vulnerability with communities, cultural objects, values, norms, places, and institutions. These attachments are constitutive of ourselves as "narrated identities" and call forth emotional, interpretive, practical, and ethical agency to "care for the future" despite its deep uncertainty.[59] Justice as recognition cannot be achieved without care for these constitutive webs of relationships. Moreover, openly recognizing the ways in which harms are done as a result of the administrative social imaginary establishes (or implies) duties of repair, reconciliation, or restorative justice. So while the social dimension of recognition creates the conditions in which the social imaginary might be reconfigured, the psychological dimension establishes the expectation that a reconfigured imaginary might be based on care.

Implications for Climate Geoengineering, and the Case for Invented Participation

Finally, in this chapter, I revisit the issue of geoengineering as a response to climate change and seek to apply the political approach to recognition. The challenges of climate change are in many respects failings of recognition. Rather than conjuring a single unified humanity, responses to climate change need to openly acknowledge and recognize diversity and difference: "our various different attitudes to risk, technology and well-being; our different ethical, ideological and political beliefs; our different interpretations of the past and our competing visions of the future."[60]

Issues of recognition also therefore arise in relation to geoengineering. Not only are forms of knowledge and ways of knowing outside conventional empirical science largely unrecognized in the modeling and technical literature that dominates climate geoengineering but so are groups of people. In common with many Anthropocene discourses, there is a tendency for geoengineering discourses to constitute a single humanity and thus a lack of recognition of difference (this extends to nonrecognition of diverse forms and degrees of vulnerability and different preferences).[61] In particular, future people are misrecognized in the dominant liberal social imaginary that frames almost all geoengineering debate.[62] And the present effects of past injustice—including colonialism, industrialization, and fossil fuel exploitation—are also typically unrecognized. All these failures of recognition deny meaningful agency to those affected.

Franziska Dübgen's critique of development aid offers helpful parallels for geoengineering as misrecognition.[63] Drawing on Fanon, Dübgen describes the "psychological violence" of dominance and subordination reproduced by misrecognition and epistemic injustice. Postcolonial misrecognition allows former colonies to "eat at the master's table," or in other words, join the global economy on neoliberal conditions and principles (enforced with structural adjustment). Agency is reserved for the postcolonial powers, and decolonization is represented as a gift, reinforcing an inferiority complex. In addition, Raewyn Connell[64] suggests, following Nandy, that colonialism also reconstructs the identities of the colonists toward militarist, nationalist technocratic and patriarchal values. This insight that misrecognition also reshapes the oppressor's identity in harmful ways can perhaps be generalized. It certainly suggests an analogy with the geoengineering literature, in which some of the proponents appear to project a paternal, colonial hubris—as virile scientists guiding and nudging childlike or feminine publics and politicians toward climate progress.[65]

Such an understanding of postcolonial center-periphery relations in terms of (mis)recognition suggests several further strong parallels with geoengineering: agency is reserved for the technocratic elite (the managers of aid projects and the high priests of climate modeling—or governments following their advice); the intervention offered becomes a way to assuage guilty consciences in the rich North (geoengineering rather than mitigation; aid rather than changing the terms of trade); and local knowledge—about how things work and are affected on the ground—is ignored or devalued (in a form of epistemic injustice).[66] For instance, the Solar Radiation Management Governance Initiative (SRMGI) primarily shares Northern-generated scientific knowledge and struggles to reach beyond Southern elites educated within the bounds of Northern traditions and theory, despite worthy goals.[67] It does not appear to foster—or even be open to—demands for recognition that extend beyond Southern interests understood through the lens of the dominant imaginary. Moreover, most geoengineering discourse (and research) seems to reflect the interests of Northern wealthy elites—the same interests that have spawned climate denialism as a means to resist emissions reductions so far.[68] The challenge of recognition, however, goes beyond the question of different preferences or interests, which Daniel Heyen et al. show can seriously distort assessments of geoengineering, but to the ways in which such preferences and associated identities may be formed.[69]

Marion Hourdequin elaborates four reasons geoengineering generates particular demands for recognition and participation: first, the global scale; second, the risks and uncertainties involved; third, the intentionality inherent in geoengineering; and fourth, the absence or inadequacy of processes for dispute settlement in this space.[70] Kyle Whyte argues that for those engaged in early geoengineering research, recognition implies actively seeking consent from Indigenous peoples with distinctive world views, noting that expert judgments of significance and urgency should not simply be permitted to override and silence dissenting views from Indigenous peoples.[71]

Such arguments hint at a move beyond the liberal social imaginary but beg the questions of whether such participatory parity enables such groups to begin to redefine the processes through which participation is achieved and through which participation leads to political change or whether, to the contrary, the participatory process works to define and reshape the groups consulted. The participatory turn in research on emerging technologies in recent years has led to much deeper and richer discussions.[72] But it is also clear that the processes cocreate new publics, often in ways that reproduce researchers' and policy makers' expectations of the technologies involved.[73] In the same way as existing power relations mold individual identities,[74] so do existing power relations shape the publics with whom researchers engage.[75]

Turning public participation into real political recognition therefore suggests a further step toward invented rather than invited forms of participation,[76] in which insurgent groups codefine not only their identities but also the processes through which they are recognized. Such processes will inevitably be iterative as identities in turn are redefined in engagement and engagements are restructured to include new identities—including those taken on and mobilized by representatives of those interests unable to directly participate. Such spaces and processes offer the potential for the emergence of care-based imaginaries, within which restorative justice may come to the fore.

Insofar as recognition genuinely offers a possibility of restorative justice for those harmed by climate change, the nexus of inequalities around fossil fuel extraction and use, the potential substitution of geoengineering for accelerated mitigation by high-emitting groups and countries, and the adaptation financed by the rich world appears especially problematic. The imaginaries of BECCS appear to have justified such a substitution already, while SAI technologies, with their relative low cost and high leverage, appear particularly vulnerable to such an effect in the future, especially if climate targets are further redefined in terms of temperature outcomes rather than carbon budgets. At present, it is generally understood that achieving a safe and stable climate requires reduced emissions in a limited carbon budget. BECCS, by delivering "negative emissions," allows temporary overshoot of carbon budgets. Worse, if the goal were to be reduced to simply achieving a particular global temperature, then humanity could—notionally—blow through any given carbon budget, continue emitting, and offset the rising temperatures by deploying SAI.

Conclusion

The implications of such a reworking of participation, politics, and social imaginaries are deep and broad. They encompass the whole of climate policy, ideas of progress, development, technology, and more. Mapping out all these implications is impossible within a short chapter. But until this process is begun, to pursue current technological imaginaries of geoengineering appears counterproductive, simultaneously sustaining the current social imaginary and its discriminatory relations and risking lock-in to a technological pathway with high risks and continued domination in the climate sphere.

Nonetheless, this analysis leads not to an outright rejection of CGE but to an understanding that the recognition of different groups, interests, and values in the climate change space would imply a transformation of institutions and politics, particularly in response to the demand for restorative justice. In turn, this might imply very different coproductions of CGE technologies or techniques (in particular, perhaps carbon dioxide removal [CDR] techniques in restorative modes). In such modes and imaginaries, agency would no longer be reserved for the technocratic elite—with new technologies actively coproduced by communities on the ground in the global periphery and imaginatively influenced by representatives of future people. Moreover, the interventions offered would no longer simply assuage guilty consciences in the rich North but be embedded in the implementation of meaningful duties of restitution (financial and otherwise) for the harms of climate colonialism.

By applying a CPE perspective to justice as recognition and its implications for geoengineering, this chapter has highlighted the ways in which the coconstruction of climate policy and geoengineering proposals reflect and promise to sustain a postpolitical (neo)liberal social imaginary. It has advocated a focus on justice as recognition as a way to expose not only the shortcomings of current climate policy (including its reliance on carbon geoengineering) but also the limitations of emerging proposals for solar geoengineering. Most importantly, it suggests that by acting on demands for recognition by the current unheard and misrecognized voices in the climate debate, we open the possibility of a transformative shift in society and the scope to develop a care-based social imaginary rooted in reconciliation and restorative justice, which could in turn help coproduce new imaginaries of geoengineering that might be deployed to such ends.

The debates over recognition, briefly discussed here, illustrate a further conclusion. The ways in which the views of both advocates and critics of recognition are shaped by the dominant social imaginary serve to undermine any emancipatory or transformative power of recognition. The scope of recognition only becomes clear from a CPE perspective that problematizes and politicizes the ways in which the dominant imaginary constructs and conditions identity and promotes both agonistic and dialogic engagement in rebuilding healthy and recognized narrative identities. Yet the CPE perspective itself is unrecognized by the dominant imaginary. Academic and technical research is almost entirely conducted at a "realist" register, with critique largely posed from a "constructivist" register. As critical scholars, we have our own struggle for recognition to pursue too.

Notes

1   Mike Hulme, *Weathered: Cultures of Climate* (London: Sage, 2017), 200.

2   Andreas Malm, *Fossil Capital: The Rise of Steam Power and the Roots of Global Warming* (London: Verso, 2016), 496; Jason W. Moore, introduction to *Anthropocene or Capitalocene? Nature, History, and the Crisis of Capitalism*, ed. Jason W. Moore (Oakland, Calif.: PM Press, 2016), 1–13; Timothy Mitchell, *Carbon Democracy: Political Power in the Age of Oil* (London: Verso, 2011), 488.

3   Christopher Groves, *Care, Uncertainty and Intergenerational Ethics* (London: Palgrave Macmillan, 2014), 251.

4   Sabine Fuss et al., "Betting on Negative Emissions," *Nature Climate Change* 4 (2014): 850–853; Glen P. Peters and Oliver Geden, "Catalysing a Political Shift from Low to Negative Carbon," *Nature Climate Change* 7 (2017): 619–621.

5   David W. Keith, *A Case for Climate Engineering* (Cambridge, Mass.: MIT Press, 2013), 224; Simon Nicholson and Michael Thompson, "To Meet the Paris Climate Goals, Do We Need to Engineer the Climate?," *The Conversation*, February 23, 2016, http:// theconversation.com/to-meet-the-paris-climate-goals-do-we-need-to-engineer-the -climate-46664; Douglas G. MacMartin, Katherine L. Ricke, and David W. Keith, "Solar Geoengineering as Part of an Overall Strategy for Meeting the 1.5°C Paris Target," *Philosophical Transactions of the Royal Society A* 376, no. 2119 (2018): 20160454.

6   Keith, *Case for Climate Engineering*, 224; Joshua Horton and David W. Keith, "Solar Geoengineering and Obligations to the Global Poor," in *Climate Justice and Geoengineering: Ethics and Policy in the Atmospheric Anthropocene*, ed. Christopher J. Preston (London: Rowman & Littlefield, 2016), 79–92.

7   Marion Hourdequin, "Justice, Recognition and Climate Change," in *Climate Justice and Geoengineering*, 33–48.

8   David Tyfield, "'What Is to Be Done?' Insights and Blind Spots from Cultural Political Economy(s)," *Journal of Critical Realism* 14, no. 5 (2014): 530–548; Markusson et al, this volume, chap. 14.

9   Duncan P. McLaren, "Mirror Mirror: Fairness and Justice in Climate Geoengineering" (PhD thesis, Lancaster University, 2017).

10 Lois McNay, *Against Recognition* (Cambridge, U.K.: Polity, 2008), 240; Paddy McQueen, "Social and Political Recognition," Internet Encyclopedia of Philosophy, accessed June 13, 2018, http://www.iep.utm.edu/recog_sp/.

11 Groves, *Care, Uncertainty and Intergenerational Ethics*, 251.

12 David W. Keith, "Geoengineering the Climate: History and Prospect," *Annual Review of Energy and the Environment* 25 (2000): 245–284.

13 Jack Stilgoe, *Experiment Earth: Responsible Innovation in Geoengineering* (London: Routledge, 2015), 240; McLaren, "Mirror Mirror"; Duncan P. McLaren, "Whose Climate and Whose Ethics? Conceptions of Justice in Solar Geoengineering Modelling," *Energy Research & Social Science* 44 (2018): 209–221. See Carton, this volume, chap. 3.

14 Groves, *Care, Uncertainty and Intergenerational Ethics*, 251.

15 Carton, this volume, chap. 3.

16 Christian Azar et al., "Carbon Capture and Storage from Fossil Fuels and Biomass—Costs and Potential Role in Stabilizing the Atmosphere," *Climatic Change* 74 (2006): 47–79.

17 Keith, *Case for Climate Engineering*; MacMartin, Ricke, and Keith, "Solar Geoengineering."

18 McLaren, "Whose Climate and Whose Ethics?"

19 The injustice of such failings of procedural justice becomes more harmful when the ideas of overall social benefit are also conceived and constructed, reflecting the interests and values of dominant and elite groups rather than those of the subaltern or Indigenous groups most often affected.

20 Whyte and Buck, this volume, chap. 5.

21 Thomas Michael Scanlon, *What We Owe to Each Other* (Cambridge, Mass.: Belknap, 1998), 432; Nancy Fraser and Axel Honneth, *Redistribution or Recognition* (London: Verso, 2003), 288.

22 Scanlon, *What We Owe Each Other*, 432; Fraser and Honneth, *Redistribution or Recognition*, 288.

23 Charles Taylor, *Multiculturalism and the Politics of Recognition* (Princeton, N.J.: Princeton University Press, 1992), 132.

24 Fraser and Honneth, *Redistribution or Recognition*, 288.

25 David Schlosberg, *Defining Environmental Justice: Theories, Movements and Nature* (Oxford: Oxford University Press, 2007), 256.

26 *Agent*, or *actor*, might be a better term than *person* here, as we may also be concerned with nonhuman species and ecological systems. But where the process of recognition is in part the

establishment of subjectivity for those whose agency is lacking or constrained, to use the term *agent* might be confusing.

27 Raewyn W. Connell, *Southern Theory: The Global Dynamics of Knowledge in Social Science* (Cambridge, U.K.: Polity, 2007), 272.

28 James Ingram, "The Subject of the Politics of Recognition: Hannah Arendt and Jacques Rancière," in *Socialité et reconnaissance*, ed. Georg W. Bertram et al. (Paris: Éditions L'Harmattan, 2006), 229–245; Jacques Rancière, "Introducing Disagreement," *Angelaki: Journal of the Theoretical Humanities* 9, no. 3 (2004): 3–9.

29 McQueen, "Social and Political Recognition."

30 Mark Lilla, "The End of Identity Liberalism," *New York Times*, November 18, 2016, http://www.nytimes.com/2016/11/20/opinion/sunday/the-end-of-identity-liberalism .html.

31 Lois McNay, *Against Recognition*, 240.

32 David Tyfield, "'King Coal Is Dead! Long Live the King!': The Paradoxes of Coal's Resurgence in the Emergence of Global Low-Carbon Societies," *Theory, Culture & Society* 31, no. 5 (2014): 59–81; Duncan P. McLaren, Nils Markusson, and David W. Tyfield, "Towards a Cultural Political Economy of Mitigation Deterrence by Greenhouse Gas Removal (GGR) Techniques" (working paper 1, AMDEG project, Lancaster University, March 2018).

33 John Asafu-Adjaye et al., "An Eco-modernist Manifesto," accessed June 13, 2018, http://www.ecomodernism.org/manifesto-english/.

34 Hourdequin, "Justice, Recognition and Climate Change"; Anne Phillips, "Recognition and the Struggle for Political Voice," in *Recognition Struggles and Social Movements: Contested Identities, Agency and Power*, ed. Barbara Hobson (Cambridge: Cambridge University Press, 2003), 263–273; Simon Thompson, "Recognition beyond the State," in *Global Justice and the Politics of Recognition*, ed. Tony Burns and Simon Thompson (London: Palgrave Macmillan, 2013), 88–107.

35 McNay, *Against Recognition*, 240; Irina Velicu and Maria Kaika, "Undoing Environmental Justice: Re-imagining Equality in the Rosia Montana Anti-mining Movement," *Geoforum* 84 (2017): 305–315.

36 McNay, *Against Recognition*, 240.

37 Velicu and Kaika, "Undoing Environmental Justice."

38 Fraser and Honneth, *Redistribution or Recognition*, 288; Thompson, "*Recognition beyond the State*."

39 Fraser and Honneth, *Redistribution or Recognition*, 288; Thompson, "*Recognition beyond the State*."

40 Fraser and Honneth, *Redistribution or Recognition*, 288.

41 Connell, *Southern Theory*, 272.

42 McLaren, "Whose Climate and Whose Ethics?"

43 Thompson, "Recognition beyond the State."

44 Virginia Held, *The Ethics of Care: Personal, Political and Global* (Oxford: Oxford University Press, 2006), 165.

45 Rahul Kumar, "Wronging Future People: A Contractualist Proposal," in *Intergenerational Justice*, ed. Axel Gosseries and Lukas H. Meyer (Oxford: Oxford University Press, 2009), 251–272.

46 Kumar, 254.

47 McNay, *Against Recognition*, 240.

48 Janna Thompson, "Identity and Obligation in a Transgenerational Polity," in *Intergenerational Justice*, 25–49; Dieter Birnbacher, "What Motivates Us to Care for the (Distant) Future?," in *Intergenerational Justice*, 273–300; Samuel Scheffler, *Death and the Afterlife* (Oxford: Oxford University Press, 2013), 224.

49 Ingram, "Politics of Recognition."

50 Ingram.

51 Schlosberg, *Defining Environmental Justice*, 256.

52 Miranda Fricker, *Epistemic Injustice: Power and the Ethics of Knowing* (Oxford: Oxford University Press, 2007), 188; James Bohman, "Domination, Epistemic Injustice and Republican Epistemology," *Social Epistemology: A Journal of Knowledge, Culture and Policy* 26, no. 2 (2012): 175–187.

53 Velicu and Kaika, "Undoing Environmental Justice."

54 Rancière, "Introducing Disagreement."

55 Duncan McLaren, "Framing Out Justice: The Post-politics of Climate Engineering Discourses," in *Climate Justice and Geoengineering: Ethics and Policy in the Atmospheric Anthropocene*, ed. Christopher J. Preston (Lanham, Md.: Rowman & Littlefield, 2016), chap. 10.

56 Alternative discourses may be possible, such as those explored by Holly Jean Buck, "The Need for Carbon Removal," *Jacobin*, July 24, 2018, http://www.jacobinmag.com/2018/07/carbon-removal-geoengineering-global-warming.

57 Estelle Ferrarese, "Vulnerability: A Concept with Which to Undo the World as It Is?," *Critical Horizons* 17, no. 2 (2016): 149–159.

58 David Schlosberg, *Defining Environmental Justice*, 256; Fraser and Honneth, *Redistribution or Recognition*, 288; Julian Agyeman, *Introducing Just Sustainabilities: Policy, Planning and Practice* (London: Zed Books, 2013), 216; Gordon Walker, *Environmental Justice: Concepts, Evidence and Politics* (London: Routledge, 2012), 272.

59 Groves, *Care, Uncertainty and Intergenerational Ethics*, 251.

60 Mike Hulme, *Why We Disagree about Climate Change: Understanding Controversy, Inaction and Opportunity* (Cambridge: Cambridge University Press, 2009), 436.

61 McLaren, "Whose Climate and Whose Ethics?"

62 Buck, "Need for Carbon Removal," provides a contrasting exception, as do several of the other chapters in this volume.

63 Franziska Dübgen, "Africa Humiliated? Misrecognition in Development Aid," *Res Publica* 18 (2012): 65–77.

64 Connell, *Southern Theory*, 272.

65 Sikka, this volume, chap. 7; York, this volume, chap. 12.

66 See Sikka, this volume, chap. 7.

67 A. Atiq Rahman et al., "Developing Countries Must Lead on Solar Geoengineering Research," *Nature* 556 (2018): 22–24.

68 McLaren, "Mirror Mirror"; McLaren, "Whose Climate and Whose Ethics?"; Surprise, this volume, chap. 13.

69 Daniel Heyen, Thilo Wiertz, and Peter Irvine, "Radiation Management Impacts: Limitations to Simple Assessments and the Role of Diverging Preferences" (working paper, Institute for Advanced Sustainability Studies, 2015).

70 Hourdequin, "Justice, Recognition and Climate Change."

71 Kyle Powys Whyte, this volume, chap. 5; Kyle Powys Whyte, "Indigenous Peoples, Solar Radiation Management, and Consent," in *Engineering the Climate: The Ethics of Solar Radiation Management*, ed. Christopher J. Preston (Lanham, Md.: Rowman & Littlefield, 2012), 65–76.

72 Jason Chilvers and Matthew Kearnes, "Science, Democracy and Emergent Publics," in *Remaking Participation: Science, Environment and Emergent Publics*, ed. Jason Chilvers and Matthew Kearnes (London: Routledge, 2016), 1–28.

73 Chilvers and Kearnes, "Science, Democracy and Emergent Publics"; Brian Wynne, "Ghosts of the Machine: Publics, Meanings and Social Science in a Time of Expert Dogma and Denial," in *Remaking Participation*, 99–120; Andy Stirling, "Transforming Power: Social Science and the Politics of Energy Choices," *Energy Research & Social Science* 1 (2014): 83–95.

74 McNay, *Against Recognition*, 240.
75 Rob Bellamy and Javier Lezaun, "Crafting a Public for Geoengineering," *Public Under-standing of Science* 26, no. 4 (2015): 402–417; Christopher Groves, review of "Remaking Participation: Science, Environment and Emergent Publics," by Jason Chilvers and Matthew Kearnes, eds., *Science as Culture* 26, no. 3 (2017): 408–412.
76 Eurig Scandrett, "Citizen Participation and Popular Education in the City," in *Big Ideas Thinkpiece for Friends of the Earth* (London: Friends of the Earth U.K., 2013), http://www.foe.co.uk/sites/default/files/downloads/citizen_participation_and.pdf.

7

An Intersectional Analysis of Geoengineering

Overlapping Oppressions
and the Demand for
Ecological Citizenship

TINA SIKKA

The rise in research on geoengineering that incorporates diversity, heterogeneity, and sociological analysis into the practice of science has been promising. For example, there is research on geoengineering that highlights issues of representation, disproportionate effects, and its roots in exploitative and/or imperialist science.[1] However, most examinations of gender, race, and class tied to the environment and climate change tend to focus on the structural violence of racism, misogyny, and poverty, with select pieces, particularly on the subject of gender, going further to address the ways in which geoengineering is problematic on the levels of design, the vision of control it perpetuates, and its roots in capitalist economic formations.[2] What is missing from the literature is an intersectional approach that applies the study of the "interaction between gender, race, and other categories of difference in individual lives, social practices, institutional arrangements, and cultural ideologies" and closely examines "the outcomes of these interactions in terms of power" relative to geoengineering specifically.[3]

In line with an intersectional approach, the understanding of how individual and group identities are products of multiple interacting social locations that include—but are not limited to—race, class, gender, religion, age, indigeneity, disability, and migration status is essential. Intersectionality illustrates how interactions take place "within interconnected systems and structures of power (e.g. laws, policies, state governments and other political and economic unions, religious institutions, media)" that, in turn, produce forms of oppression and privilege "shaped by colonialism, imperialism, racism, homophobia, and patriarchy."[4]

The danger of not studying the effects of geoengineering in this way opens the door to depoliticizing the suite of technologies, processes, techniques, and understandings that constitute it while also making it difficult to discern the ways in which geoengineering has the capacity to impact the lives of people and communities in unanticipated ways. Generally, assessments of geoengineering and risk tend to homogenize and/or fix identity by treating it as a monolith. Even studies of geoengineering that begin from a place of skepticism and critique oftentimes fail to identify and humanize those likely to be negatively affected. Moreover, environmental effects are frequently siloed from the human ones—with the former given precedence over the latter, which, because of the way we structure knowledge, externalizes the effects on nature and the environment in ways that undermine the severity of potential impacts. This occurs in scientific literature as well as media discussions—the latter of which is concretely responsible for shaping public perceptions.[5]

With respect to intersectionality, the discursively produced and structurally enforced subject positions of gender, race, and class discussed in this chapter are by no means exhaustive. Ethnicity, sexuality, bodily comportment, disability, age, and other categories also shape the felt effects of large-scale technological interventions like geoengineering. The traditional privileging of white male subjectivities as default is reflected in contemporary discourse around geoengineering, while the populations that will likely suffer its side effects are nameless and faceless. What I will highlight in this chapter is the way in which the categories of gender, race, and class interact with each other to produce divergent experiences of power and powerlessness based on political authority and voice, access to knowledge and resources, and levels of socioeconomic vulnerability.[6]

Overall, this chapter aims to shape future scientific and social scientific research on geoengineering such that it begins to incorporate intersectional analysis as part of its practice. This is all in pursuit of knowledge that seeks to consider grounded experiences of inequality over and above generalizations of risk and statistical potentialities. Next, I take up and discuss the categories of gender, race, and class as they relate to climate change and geoengineering before engaging in a grounded intersectional analysis of a hypothetical case study that integrates all three categories as they will likely manifest under stratospheric solar radiation management (SRM). I then conclude with a discussion of what should be done going forward.

Gender

Feminism itself is constituted by multiple feminisms—each of which has something to offer to the study of geoengineering. Here I focus on two: ecofeminism and standpoint theory.

Ecofeminism is perhaps the most directly relevant feminist approach with which to examine geoengineering because of its focus on the relationship between gender, technology, and the environment. Ecofeminists like Carolyn Merchant, Vandana Shiva, Maria Mies, and Joni Seager assert that modern technology is in and of itself patriarchal and maintain that its roots in a legacy of exploitative Western capitalist scientific and technological practices simultaneously degrade and exploit nature and women because of its orientation toward industrialization, control, and profit.[7] This is known as the "twin subordinations" thesis popularized by Françoise d'Eaubonne in 1974.

Consequently, one of the most significant contributions of ecofeminism to the study of geoengineering is its focus on the negative impacts that technologies that seek to transform natural processes for human gain have had on the lives of women.

In relation to geoengineering, ecofeminists who have discussed its testing and possible use have roundly rejected it. Shiva, for example, characterizes geoengineering as "the ultimate hubris" and states that climate change is itself the product of the same paradigm of knowledge that has given rise to geoengineering as a possible solution:

> We could replace people with fossil fuels, have higher and higher levels of indus-
> trialization, of agriculture, of production, without thinking of the green-house
> gases we were emitting, and climate change is really the pollution of the engineer-
> ing paradigm, when fossil fuels drove industrialism. To now offer that same mind-
> set as a solution is to not take seriously what Einstein said: that you can't solve the
> problems by using the same mindset that caused them. So, the idea of engineering
> is an idea of mastery. And today the role that we are being asked to play is a role
> based on informed humanity.[8]

Greta Gard takes a similar position to Shiva's and asserts that "climate change and first world overconsumption are produced by masculinist ideology, and will not be solved by masculinist techno-science approaches."[9] Claudia von Werlhof and Ana Isla, in their text *Mother Earth under Threat*, conclude that geoengineering represents a dangerous step toward planetary manipulation of the earth and all her living systems.[10]

In opposition to geoengineering, the ecofeminist solution to climate disruption is to (1) swap traditional epistemologies for ones that highlight the voices of marginalized women; (2) (re)discover modes of knowledge that are plural, diverse, and contextual; and (3) adopt a not-for-profit-oriented model of growth and development that recognizes that life in nature is maintained cooperation, interdependence, and mutual care.

Standpoint feminist theory offers several additional insights on gender and geoengineering as it relates to representation, situated knowledge, and scientific practice by directly addressing how science is conducted. It also avoids essentializing gender differences as natural or innate and treats objectivity as both epistemologically significant and ontologically possible. Standpoint theory, briefly, is a postpositivist mode of feminist thinking that is informed by critical theory, interpretivism, and Marxism.[11] Significantly, it values knowledge produced from the ground up, eschews the removal of norms from scientific practice by arguing that science is always value-laden, and proposes the formulation of standards that are maximally oriented toward justice, equity, and fairness. As Harding argues, it is the "experiences and lives of marginalized people, as they understand them . . . [that] . . . provide particularly significant problems to be explained or research agendas" often ignored by mainstream science.[12]

Standpoint theory offers a robust accounting of representation in scientific practice, which is critical, since without representation, it is difficult for non-hegemonic perspectives and practices to be heard, tried, and tested. Currently, the percentage of women working in the area of climate science is itself low, with the most recent statistics reflecting that only 20 percent of geoscientists and meteorologists are women. While they are better placed in ecology, women are not well represented in research universities or academic journals overall.[13] Funding for large research projects related to climate change is also male-dominated—as well as geographically and racially unrepresentative.[14]

The number of women working in the domain of geoengineering is also sparse. In a recent meta-analysis of author networks examining publications on geoengineering, top spots all went to men (Ken Caldeira, Bala Govindasamy, Alan Robock, and Ben Kravitz lead in this study).[15] Dr. Naomi Vaughn of the Tyndall Centre for Climate Change Research at the University of East Anglia is one of a handful of exceptions. Not having women's voices represented in geoengineering research means that the life experiences and concerns of over 50 percent of the population is more likely to be deprioritized. This exclusion explains why risks associated with climate engineering often ignore gendered specificities rooted in the lives of women. Additionally, consistent with standpoint theory, this lacunae also undermines "the potential for 'breakthrough conceptualizations [, which is] is decreased'" thus hampering the "invigorating creative tension between scientific perspectives."[16]

In addition to representation, standpoint theory's contribution to women's experience is also rooted in its focus on women's roles in both production and social reproduction, which grounds the study of environmental risk, climate change, and geoengineering in the material realities of women's lives. This perspective reflects women's responsibilities for social reproduction in the family as well as their role in so-called productive work, which is considered as important as the former, since "renewing life is a form of work, a kind of production, is fundamental to the perpetuation of society as the production of things."[17] This care work is intensified under conditions of environmental duress where access to food, water, health care, and a safe environment are jeopardized.[18] While studies of climate change have begun to examine its effects in more granular ways that include gender, race, and class, research on geoengineering has remained firmly wedded to a strategy of routine delineation of risk in the form of generalities.

Finally, in addition to representation, burdens, and felt effects, standpoint theory also highlights the kinds of epistemic injustices that can occur when one kind of scientific practice is privileged over others and, in contradistinction to ecofeminism, offers an alternative that is likely more actionable than ecofeminism's radical approach. According to Sandra Harding, standpoint theory, in privileging marginalized voices and incorporating just social values, constitutes the practice of "good science."[19] What makes this approach particularly useful as it relates to geoengineering is that it makes a strong case for the inclusion of women's voices in science as essential to the maintenance of sound scientific practice and raises questions about research paradigms that value power, control, and abstraction in a way that resists essentializing femininity.[20] Scientific literature on geoengineering related to prospective modeling, for example, is conspicuous in its use of language and assumptions that rely on this maximally technoscientific framework. Standpoint theory would question the hegemony of rooting geoengineering models in a "model of predictive control" framework using "standardized experimental conditions" that considers success solely in relation to operational effectiveness.[21] Conversely, it advocates for a practice of science that relativizes—but is not relativist—by asking questions from the margins, including multiple perspectives and methods, and articulating novel knowledge projects that, with respect to geoengineering and climate change, would challenge its operating logic(s).[22]

In closing this section, it is important to acknowledge the important contribution made by two articles that address the gendered implications of geoengineering directly. The first, by Diane Bronson, addresses the problems associated with the scarcity of women's voices in geoengineering research and examines the language of geoengineering, which she argues is "explicitly gendered and filled with sexual imagery."[23] The second piece provides an overview of the overwhelmingly male demographics of geoengineering scientists as well as its advocates, its masculinist ideology of control, power, and instrumentality, and the ways in which the design of geoengineering itself "consists of the simultaneous play and display of technical prowess in the masculine 'ritual of tinkering.'"[24]

Having discussed gender at length, the next sections take up race and class before engaging in an intersectional analysis that dereifies the boundaries between identity markers using a grounded case study.

Race

Race represents a significant marker of identity that has not been adequately represented in environmental science generally or climate engineering science in particular. The scientific and cultural fault to white subjectivity coupled with environmental racism, wherein racialized minorities are made to bear the brunt of environmental degradation, are two of the primary reasons this is the case. Race, in this context, is understood as a sociohistorically produced set of categories made manifest through discourse, policies, social interactions, and norms that have materially, bodily, and psychically significant consequences.[25]

Issues of representation are as significant to race as they are to gender, class, disability, sexuality, and other forms of marginalization. Scientists working on geoengineering tend to be as unrepresentative with respect to race as in other STEM fields and in the environmental sciences specifically. In a comprehensive census of the 66,502 environmental scientists and geoscientists in the United States in 2011, a dismal 5 percent were Asian, 2.7 percent were Black, and 3.9 percent were Hispanic.[26] While there is no data on the demographic breakdown of climate engineering scientists, a cursory look at the board and review panel of the Royal Society geoengineering working group and other contributors to large-scale reviews of geoengineering reveal low levels of racial and ethnic representativeness.[27] This lack of representation is a tangible manifestation of structural inequities and outright discrimination that have a long history in the way science is practiced—that is, in ways that it ignores marginalized voices—has been used to justify racist practices and, with respect to the environment, leads to decisions and regulations that disproportionately and negatively impact groups who have little decision-making or knowledge-producing power.[28] It is also illustrative of how this silencing can result in the replication of racial privilege and the excising of alternative perspectives that can result in more innovative science.

The cumulative effects of a lack of representativeness, from the perspective of race, is aptly synthesized by Harding, who, as mentioned in the previous section, makes a case for the inclusion of epistemologies that are non-Eurocentric, reflective of local concerns, and rooted in the lived experiences of racialized communities.[29] Specifically, Harding examines the issue of representation through a postcolonial lens and argues for the centering of marginalized epistemologies that might "make important contributions to the storehouse of global scientific knowledge."[30]

With respect to geoengineering, this would include not only participation in the front end of geoengineering research but a kind of participation that is formative in ways that would likely derail or, at the very least, transform how climate change itself is conceived of and how solutions are reached. This, as a result, would challenge the idea that issues related to environmental racism and postcolonialism are peripheral.[31] While the subject of racial representativeness in geoengineering research is significant and deserves a more rigorous discussion, an equally consequential set of arguments articulating the intersections between climate change, geoengineering, and race revolves around the discourses and material consequences of environmental racism.

A significant body of literature on race and the environment pertinent to geoengineering focuses on environmental racism defined (loosely) as "racial discrimination in environmental policy-making and enforcement of regulations and laws; the deliberate targeting of communities of color for toxic-waste facilities; the official sanctioning of the presence of life-threatening poisons and pollutants in communities of color; and the history of excluding people of color from leadership in the environmental movement."[32]

The experience of environmental racism in marginalized communities ranges from the troublingly quotidian to the dire and immediately life-threatening: "Millions of African Americans, Latinos, Asians, Pacific Islanders, and Native Americans are trapped in polluted environments because of their race and color. Inhabitants of these communities are exposed to greater health and environmental risks than is the general population. Clearly, all Americans do not have the same opportunities to breathe clean air, drink clean water, enjoy clean parks and playgrounds, or work in a clean, safe environment. People of color bear the brunt of the nation's pollution problem."[33]

This has been statistically mapped by Bullard, and supported by others, who documents how "race has been found to be an independent fact, not reducible to class, in predicting the distribution of air pollution . . . contaminated fish consumption . . . the location of municipal landfills and incinerators . . . the location of abandoned toxic waste dumps . . . and lead poisoning in children."[34] Environmental racism, as a field of research, is complex and makes the case that ecological and racial injustices are coproduced and that decision-making in a capitalist economic system "reinforces and reproduces the dominance of the basic structures that are behind the generation of the environmental crisis—which are the structures behind its own [racism's] generation."[35]

It is important to remember that this framework, like most contemporary studies of race, treats racism less as a set of discrete acts by individual people and more as an outcome of structures rooted in particular spaces and places.[36] Putting this basic insight into practice in her own work, Pulido adds that in the case of the environment, it is how relationships *between* places produce racism that has been overlooked. While Pulido focuses on correlations between the spaces that constitute urban environments, particularly with respect to the racialization of different zones in cities, she also examines the relationship between these zones to rapid "industrialization, decentralization, and residential segregation" and the correlation between specific spaces, high levels of pollution, and race.[37]

On the level of crude economics, it is also the case that land located next to racialized communities tends to be cheaper, and thus polluting industries, sewage plants, incinerators, and so on are more likely to be located nearby. This indicates a strong "correlation between race, income and residence [which] influence several outcomes such as a higher likelihood of being exposed to environmental hazards, the disproportionate impacts of environmental processes and policies, the targeting and siting of noxious facilities in more deprived communities and inequalities in the delivery of environmental services such as rubbish removal."[38]

Consequently, the places in which geoengineering is likely to take place is also of critical significance. There is a history of documented evidence of waste disposal, hazardous sites, and industrial activity being housed in and next to vulnerable and disempowered communities.[39] Although geoengineering's effects, whether SRM or carbon dioxide removal (CDR), are likely to be globally felt, based on this history of environmental injustice, we can extrapolate that the most extreme impacts of climate engineering will likely be felt by racialized communities where vulnerabilities are most strongly experienced.

The lack of tangible research examining the role of race in shaping the trajectory of geoengineering or factoring it in as a significant variable in any assessment of effects is striking. The disproportionately felt effects of, for instance, aerosol spraying would likely be experienced by racialized communities that would be made vulnerable by ozone depletion, a lack of resources to cope, diminished rainfall, blocked sun (leading to vitamin deficiencies), and changes to agriculture.[40]

A similar set of arguments that capture the intersection of race, place, and the environment can also be made in a more geographically global context using work done on postcolonialism. In accordance with this approach, it is important to identify how environmental disasters in countries of the Global South have been historically handled, or mishandled, in the past. Such environmental disasters as the Union Carbide disaster in Bhopal, India, and the state of the oil industry in the Niger delta, whose population has suffered as a result of environmental accidents linked to petroleum extraction, set a troubling precedent for how accidents associated with geoengineering might be handled in the future.[41] The insertion of transnational corporations into this equation further reveals the material effects of unequal power relations on a global scale, whose effects will be most acutely felt by marginalized Others. The racialization of space and its maintenance through state and corporate power is a constant in both cases.

A particularly significant example of how race, postcolonialism, and the environment intersect with geoengineering is in the case of a nongovernment-sanctioned iron fertilization test that took place in 2012 in northern British Columbia, Canada, in the waters off Haida Gwaii. Briefly, in 2012, Russ George (an American entrepreneur and vocal advocate of geoengineering) persuaded the heads of the Council of the Haida Nation to form the Haida Salmon Restoration Corporation (HSRC), which dumped one hundred tons of iron into the Pacific Ocean as part of what amounted to an unsanctioned geoengineering experiment. The expedition, which George partially funded, was ostensibly aimed at reinvigorating lagging salmon stocks by increasing phytoplankton blooms as well as paving the way for the Nation to profit off captured carbon through carbon credits.[42] However, the test was found to have contravened moratoria on iron fertilization set by the U.N. Convention on Biological Diversity (CBD) as well as the London Convention's ban on dumping.[43] In the end, criminal investigations were opened, and there was little evidence that the experiment resulted in either increased salmon runs or a monetizable capture of carbon.[44]

The relevance of this case in the context of postcolonialism, race, and geoengineering is twofold. First is the way in which the community's economic and ecological stress was exploited for private gain in a manner reminiscent of past colonial forms of resource exploitation and control.[45] The significance of salmon to the Haida Nation in the form of food, art, and storytelling is critical since for "many Haida people, protecting and securing access to salmon is fundamental to reclamation of the Haida cultural identity and autonomy, after deliberate and systematic colonial violation of the Haida way of life."[46]

The second critique focuses on the way in which this very fact of historical oppression, based on racialization and resource exploitation, was reflexively stitched into the corporate media narrative that asserted that this form of geoengineering represented a way for the community to reassert sovereignty over their lands. The company's website, for example, asserted that the project represents "world class science, village style," with John Disney, the head of the company, claiming that "it seemed appropriate for Old Massett to take the first steps to reclaim their stewardship role by working in an area that, before contact, would have been their responsibility."[47] This form of greenwashing co-opts the Indigenous discourse of environmental resilience and respect while also rendering invisible traditional knowledge and local perspectives.[48]

The very fact that this test took place in an economically precarious location, under unequal conditions of power, and between a well-funded corporate entity promising environmental rejuvenation and a historically marginalized population renders the case an important precedent in assessing the role race is likely to play with respect to the location of future geoengineering experiments as well as the level of acceptable risk and effects on the immediately subjected and affected communities.

Cumulatively, this discussion of geoengineering, climate change, and race highlights the inextricable connection between environmental injustice and race based on historical discrimination. These experiences are a result of unrepresentative decision-making about where environmental waste facilities are located coupled with a "history of excluding people of color from leadership."[49] The current dearth of discussions that center these absences and concerns in contemporary discussions of geoengineering is particularly worrying, as highlighted by the experience of the Haida in Old Massett.

Class

It should be noted that most geoengineering advocates do acknowledge the potential for unintended consequences but do so in ways that avoid specificity. This is significant, since place attachment, local messages, personalization, and emotional connection are important elements of climate communication that help the public understand the gravity of climate change and that should, as a result, form a key part of the informational matrix that surrounds geoengineering.[50]

Much like with race, it is also true that populations living in conditions of relative and absolute poverty tend to suffer more acutely from the effects of environmental disasters because of their geographical location (often situated close to the most environmentally vulnerable areas), their general lack of economic resources, and an absence of political power and representation tied to class status. Susceptibility to environmental vulnerabilities is made worse by climate change, which is itself tightly correlated to class-specific forms of poverty.[51]

Moreover, poorer populations tend to be the most at risk in relation to pollution, waste disposal, waste containment, and natural disasters. These communities, on a global level, are also likely to house some of the most loosely regulated and highly polluting industries situated in close proximity to homes, schools, and sources of food and water.[52]

These economically marginalized groups have also had to manage the most extreme consequences of climate change, whose distributional effects are heavily skewed toward the poor. With respect to geoengineering, if we follow the precedent set out previously, the class-based consequences of deploying these technologies are likely to follow this same pattern, wherein tests, trials, and deployments will take place in marginalized, less-well-regulated spaces and whose effects—whether they are stratospheric sulfate–linked ozone depletion, regional changes in surface temperature, or changes to the water cycle—will be felt the most by the least materially well off. Arguments that contend that the poor might benefit from geoengineering have a considerable amount of work to do in order to meet a high bar of evidence and address the risks noted above.

Another particularly salient set of arguments regarding geoengineering and its relationship to class has to do with its characterization as an ill-conceived solution to a social problem rooted in modern capitalism. Following this line of thinking, it is argued that neoliberal capitalism, in its current manifestation, is itself to blame for climate change because of its privatized, extractive, and profit-oriented ethos that privileges accumulation and growth at the expense of democratic control and equity and that creates the permission structure for intensified CO_2 emissions to get to the levels that they have.[53]

A focus on equity, sustainability, social cohesion, and prosocial environmental policies as the only just path by which to address climate change emphasizes the significance of inequities of class as a particularly burdensome outcome of neoliberal capitalism. Geoengineering can be seen as an extension of this same neoliberal logic in that it perpetuates a capitalist mode of production and growth.

Cumulatively, assessing climate engineering through the lens of class using existing research on the effects of environmental disasters and climate change as precedents paints a discouraging view of its potential consequences. As with gender and race, class or socioeconomic status forms another line of discrimination that contemporary studies of geoengineering have failed to adequately reflect. Having access to certain levels of class privilege, particularly within a neoliberal economic system, means that access to a safe and liveable environment with a semblance of financial security, in spite of climate change and the effects of geoengineering, is far from assured for much of the world's population.

Intersectionality: A Case Study

Intersectionality has been used to examine the consequences of climate change by a handful of scholars, including Kaijser and Kronsell, whose work examines the ways in which differential power relations—based on group identity—are reflected in climate change policies and structures, and Natalie Osborne, who offers a way in which to integrate intersectional analysis into contemporary adaptation work.[54] Further research that examines the underrepresentation of marginalized groups in scientific decision-making, the vulnerability and injustices experienced by these groups as a result of climate disruptions, and the material consequences of unequal power relations has also been used to demonstrate how an individual's social location shapes their experience of climate change.[55]

Given this and the preceding insights into how gender, race, and class shape experiences of climate change and geoengineering, I argue that it is important to think about the ways in which these categories are likely to overlap and intensify vulnerability and marginalization with respect to particular forms of geoengineering. If we take the categories of gender, race, and class, keeping in mind these are not exclusive markers of identity, to construct an account of the embodied experience of a marginalized person affected by the unanticipated consequences of geoengineering (in this case, stratospheric SRM), a more granular and felt sense of risk, ethics, and responsibility is possible.[56]

When scientific studies of stratospheric SRM do acknowledge that unequally felt consequences of geoengineering are likely, they often do so in generalities—for example, by pointing out that "geoengineering impacts will differ substantially in socio-economic terms according to the relationship between climatic change and for example, distributions of population and cropland"; that its abrupt termination will likely hasten warming; or that "SRM . . . could affect food security through effects on marine food webs."[57] This research does not do an adequate job of discussing what these impacts *will look and feel like* for marginalized groups.[58] It is the felt and embodied effects experienced by individuals who suffer multiple forms of oppression based on interrelated identity markers that geoengineering research has thus far not accounted for.[59]

We now turn to the central case study, which assesses the danger that stratospheric SRM schemes could lead to reductions in the intensity of monsoons in regions of Africa as well as in Southeast Asia.[60] Leading scholars have identified knock-on effects of a reduction in precipitation as potentially leading to food insecurity, manifested in a lack of consumable food and potable water for upward of two billion people; population dislocation; and political instability.[61] In the abstract, a simple statement of risk in scientific studies does meet the criteria of responsible scientific practice in that it articulates that the danger is location specific and demonstrates a concern for vulnerable populations. However, I argue that this approach is wholly insufficient.

What, for instance, would drought-like conditions brought on by stratospheric geoengineering mean for a thirtysomething, married, lower-caste (or "scheduled" caste) Indian woman who has four children from a small village in rural Uttar Pradesh (UP)? To be clear, while it may not be possible to assess multiple permutations of this kind all the time, these exercises are important, as they encourage a more dynamic and grounded sense of how revolutionary technologies like geoengineering could perpetuate existing systems of oppression or, as Dhamoon puts it, "differing degrees and forms of penalty and privilege."[62]

By way of some context, UP is a poor, largely agricultural state located in northern India with a population of over two hundred million.[63] There have already been documented cases of climate extremes, including flooding and drought in the state, which have been devastating for its largely rural population of small-scale farmers whose major crops include cereals, pulses, and sugarcane. These climatic fluctuations are particularly significant, since a successful harvest is heavily dependent on the June to September monsoon season.[64]

Using an intersectional framework informed by the preceding discussions, on the level of gender, the burdens that would be visited upon our hypothetical woman as a consequence of geoengineering could be overwhelming. Women in UP live in conditions that are shaped by low levels of literacy, a heavily skewed sex ratio (as a result of a historically complicated preference for male children), high maternal mortality rates, and attenuated power in decision-making.[65]

Moreover, it is likely that this woman would have significant responsibilities for care work in the home, including for children and extended family. Moreover, as my discussion of ecofeminism and standpoint theory makes clear, it is often the case that when the environment is compromised, it is women and their dependents whose vulnerability is multiplied. Having an inadequate capacity to deal with the "shocks and stress[es] to which an individual or household is subject" as well as with "an internal side which is defencelessness, meaning a lack of means to cope without damaging loss," is both psychically and physically challenging.[66]

Ecofeminism's emphasis on the coextensive relationship between women and the natural world, however constructed this relationship may be, further exposes burdens under geoengineering from the perspective of environmental resource management and basic agricultural work, since women are more likely to be responsible for "preparing the ground, sowing, weeding, growing vegetables, harvesting, threshing, winnowing, drying, boiling (mainly paddy) and storage, etc."[67] Drought-like conditions triggered by geoengineering might also require the physical movement of entire families to more stable sources of sustenance—which itself is demanding for women as well as dangerous because of the high risk of rape and sexual assault while both in transit and in refugee camps.[68]

The effects of racialization in the context of the risks associated with stratospheric geoengineering for this hypothetical woman could express itself in two ways. On a global level, there is both an implicit and explicit mani-festation of racism connected with how rich nations of the Global North, both governments and individuals, deal with countries and populations in distress—namely, with a mixture of insufficient assistance immediately after an event coupled with neglect thereafter. There is strong evidence to suggest that there is a racial dynamic to aid distribution, which tends to fluctuate as a result of media framing and popular sentiment.[69] It is unlikely that this pattern will change in the aftermath of unintended environmental effects associated with geoengineering.

Moreover, media coverage of nonwhite bodies in distress—whether it is as a result of natural disasters, terrorism, or climate change, as precedent would have it—is likely to lead to identity negation, where the displaced are treated like an undifferentiated mass; compassion fatigue, in which the public, in light of spectacularized and repeated images of suffering, are left emotionally exhausted and inured to the misery they see; and a rise in populist and racist politics.[70] Additionally, a likely unfortunate outcome of geoengineering gone wrong is that displaced individuals are likely to continue to lack voice and agency in the mechanics of postdisaster relief and recovery—particularly when movement is involved.[71]

It is also important to ask if in fact this woman is forced to migrate outside of her native country, would she and her family be considered migrants? Refugees? Climate refugees? And what rights and supports would be extended to her? As the public, media, and governmental reactions to the current refugee crisis has made clear, this woman and her family could as easily be met with hostility and antagonism rather than openness and care.[72]

On a local level, within a country like India, it is also probable that our hypo-thetical woman, coming from a lower caste, would be living in parts of UP that are already vulnerable to drought due to the political nature of the way in which land is distributed.[73] Moreover, as a result of the social constraints placed on caste, which is tightly tied to occupational possibilities, it is often difficult for farmers to engage in other forms of economic activity.[74] Added to these barriers, recent evidence examining how relief is distributed postdisaster, using the 2004 Indian Ocean tsunami recovery as a case study, found that villages with inhabit-ants that were "minorities, outcastes, and non-members . . . were often excluded from the assistance process," making the capacity for recovery insufficient.[75]

Moreover, because the majority of farms in UP are in the form of small-holdings, the class dynamics of affected farmers and women in particular is especially stark. For women, even without the added risk of geoengineering-induced climate disruption, family finances are often insecure and tied to access to a stable food supply. Food insecurity has been already exacerbated by a changing climate that has been devastating for farmers and their families.[76]

Finally, it is important to emphasize that the class-related implications of climate engineering for this hypothetical woman are tightly tied into her racialized caste status and gender as well as shaped by the experience of poverty. In UP, the average monthly income for farmers is just 4,923 rupees a month.[77] High levels of indebtedness to moneylenders as a result climate change–related crop failures and overall lack of social support has led to a raft of suicides among farmers, leaving women in extreme crisis and carrying large debts.[78] Moreover, the education of children is often impossible in these circumstances, leading to a continued cycle of poverty for subsequent generations, an increase in the rates of child labor, and particularly for girls, marriage at a young age.

As has been established, this woman's class status makes it more likely that she will lack the resources needed to cope with drought exacerbated by climate engineering in terms of material resources and social support.[79] Furthermore, her family's access to and ownership over land is liable to be tenuous, leading to difficulty in attaining social mobility. Finally, it is important to note that these socioeconomic conditions have to be seen in the context of neoliberalism and under conditions of large-scale resource extraction, infrastructure projects, and state and corporate land grabs that render much of the land in UP polluted, limited in size, and of low quality.[80]

Overall, if we draw out and matricize these potential effects of stratospheric sulfate geoengineering in a manner that eschews generalities for particulars and does so from an intersectional perspective that sees power, privilege, and oppression as outcomes of classed, gendered, and racialized hierarchies, it becomes clear that contemporary studies of geoengineering have a considerable amount of work to do in engaging with risk, responsibility, and equity on a more granular level. As has been argued, current research on SRM geoengineering has not sufficiently incorporated significant social factors into scientific practice, including modeling, scenario building, and/or small-scale trials. This has led to an understanding of geoengineering and its effects that eschews the impact it might have on marginalized groups, particularly on those that experience multiple and overlapping forms of oppression based on—but not limited to—their gender, race, and class, for impacts that are generalized and centered on privileged populations.

The consequences of this aporia are demonstrated concretely in the hypothetical scenario in which this lower-caste and rural Indian woman's outside status (from the vantage of gender), race, and class are used to make the case for a more just, inclusive, and reflexive scientific practice. Understanding how all these axes of marginalization change and morph with new technologies inclusive of geoengineering is the first step toward reaching a more robust understanding of a whole host of sociotechnical transformations currently underway.

Notes

1   Laura Mamo and Jennifer R. Fishman, "Why Justice? Introduction to the Special Issue on Entanglements of Science, Ethics, and Justice," *Science, Technology, & Human Values* 38,

no. 2 (2013): 159–175; Duncan P. McLaren, "Mirror Mirror: Fairness and Justice in Climate Geoengineering" (PhD thesis, Lancaster University, 2017); Jane A. Flegal and Aarti Gupta, "Evoking Equity as a Rationale for Solar Geoengineering Research? Scrutinizing Emerging Expert Visions of Equity," *International Environmental Agreements: Politics, Law and Economics* 18, no. 1 (2018): 45–61.

2 Kevin O'Brien, "First Be Reconciled: The Priority of Repentance in the Climate Engineering Debate," in *Theological and Ethical Perspectives on Climate Engineering: Calming the Storm*, ed. Thomas Bruhn et al. (Lexington, Md.: Lexington Books, 2016), 197–203; David Russell Zeller, "There Is No Planet B: Frame Disputes within the Environmental Movement over Geoengineering" (PhD diss., University of South Florida, 2017); Holly Jean Buck, Andrea R. Gammon, and Christopher J. Preston, "Gender and Geoengineering," *Hypatia* 29, no. 3 (2014); Diana Bronson, "Geoengineering: A Gender Issue?," *Women in Action* 2 (2009); Tina Sikka, "A Critical Discourse Analysis of Geoengineering Advocacy," *Critical Discourse Studies* 9, no. 2 (2012).

3 Inez Torres Davis, "Environmental Oppression" (lecture, *ELCA's 2003 Caring for Creation New Consultation*, University of St. Mary of the Lake, Mundelein, Ill., 2003), http://www.webofcreation.org/lens/13-resources/28-lecture-on-environmental-racism-by-inez-davis; Leslie McCall, "The Complexity of Intersectionality," in *Intersectionality and Beyond: Law, Power and the Politics of Location*, ed. Emily Grabham et al. (London: Routledge-Cavendish, 2008), 65–92; Ange-Marie Hancock, *Intersectionality: An Intellectual History* (Oxford: Oxford University Press, 2016).

4 Olena Hankivsky, *Intersectionality 101* (Burnaby, B.C.: Institute for Intersectionality Research & Policy, Simon Fraser University, 2014), 2.

5 This is in regard to geoengineering specifically. Geoengineering is treated in this article less as individual technologies than as a suite of techniques with common characteristics or as an "ideal type." In essence, "An ideal type is formed by the one-sided accentuation of one or more points of view and by the synthesis of a great many diffuse, discrete, more or less present and occasionally absent concrete individual phenomena, which are arranged according to those one-sidedly emphasized viewpoints into a unified analytical construct." Max Weber, *Max Weber on the Methodology of the Social Sciences* (New York: Free Press, 1949), 90. The objective of the ideal type construct as an analytical tool is not to generalize but to treat a set of complicated historical phenomena or, in this case technologies, as having a particularly unique set of discernable and significant characteristics that are of particular importance. Matti Luokkanen, Suvi Huttunen, and Mikael Hildén, "Geoengineering, News Media and Metaphors: Framing the Controversial," *Public Understanding of Science* 23, no. 8 (2014): 966–981; Samantha Scholte, Eleftheria Vasileiadou, and Arthur C. Petersen, "Opening Up the Societal Debate on Climate Engineering: How Newspaper Frames Are Changing," *Journal of Integrative Environmental Sciences* 10, no. 1 (2013): 1–16.

6 Natalie Osborne, "Intersectionality and Kyriarchy: A Framework for Approaching Power and Social Justice in Planning and Climate Change Adaptation," *Planning Theory* 14, no. 2 (2015): 130–151.

7 Vandana Shiva, "Development, Ecology and Women," in *Earthcare: An Anthology in Environmental Ethics*, ed. David Clowney and Patricia Mosto (Lanham, Md.: Rowman & Littlefield, 2009), 274; Vandana Shiva and Maria Mies, *Ecofeminism* (London: Zed Books, 2014); Joni Seager, *Earth Follies: Feminism, Politics and the Environment*, vol. 11 (London: Routledge, 2019).

8 Vandana Shiva, "Terra Futura 2013: Interview with Vandana Shiva about Geoengineering," interview by Maria Heibel, *NoGeoingegneria*, July 9, 2013, http://www.nogeoingegneria.com/interviste/terra-futura-2013-interview-with-vandana-shiva-about-geoengineering/.

9 Greta Gaard, "Ecofeminism Revisited: Rejecting Essentialism and Re-placing Species in a Material Feminist Environmentalism," *Feminist Formations* 23, no. 2 (2011): 20.

10   Ana Isla and Claudia von Werlhof, *Mother Earth under Threat: Ecofeminism, the Land Question, and Bioengineering* (Toronto: Inanna Publications & Education, 2017).

11   Susan Hekman, "Truth and Method: Feminist Standpoint Theory Revisited," *Signs: Journal of Women in Culture and Society* 22, no. 2 (1997): 341–365; Sandra G. Harding, "Rethinking Standpoint Epistemology: What Is 'Strong Objectivity'?,'" *Centennial Review* 36, no. 3 (1992): 437–470; Sandra G. Harding, "Is Science Multicultural? Challenges, Resources, Opportunities, Uncertainties," *Configurations* 2, no. 2 (1994): 301–330; Linda Alcoff and Elizabeth Potter, *Feminist Epistemologies* (New York: Routledge, 2013).

12   Harding, "Rethinking Standpoint Epistemology.'"

13   Christopher Beck et al., "Diversity at 100: Women and Underrepresented Minorities in the ESA," *Frontiers in Ecology and the Environment* 12, no. 8 (2014): 434–436; Stephen J. Ceci and Wendy M. Williams, "Understanding Current Causes of Women's Underrepresentation in Science," *Proceedings of the National Academy of Sciences* 108, no. 8 (2011): 3157–3162; Martin Lukacs, "World's Biggest Geoengineering Experiment 'Violates' UN Rules," *Guardian*, October 15, 2012, http://www.theguardian.com/environment/2012/oct/15/pacific-iron-fertilisation-geoengineering.

14   UNESCO, "Gender and Science," Natural Sciences Priority Areas, last modified 2017, http://www.unesco.org/new/en/natural-sciences/priority-areas/gender-and-science/.

15   Paul Oldham et al., "Mapping the Landscape of Climate Engineering," *Philosophical Transactions of the Royal Society A: Mathematical, Physical and Engineering Sciences* 372, no. 2031 (2014): 20140065.

16   Rebecca Campbell and Sharon M. Wasco, "Feminist Approaches to Social Science: Epistemological and Methodological Tenets," *American Journal of Community Psychology* 28, no. 6 (2000): 778.

17   Barbara Laslett and Johanna Brenner, "Gender and Social Reproduction: Historical Perspectives," *Annual Review of Sociology* 15, no. 1 (1989): 383.

18   Sherilyn MacGregor, "'Gender and Climate Change': From Impacts to Discourses," *Journal of the Indian Ocean Region* 6, no. 2 (2010): 223–238; Justina Demetriades and Emily Esplen, "The Gender Dimensions of Poverty and Climate Change Adaptation," in *Social Dimensions of Climate Change: Equity and Vulnerability in a Warming World*, ed. Robin Mearns and Andrew Norton (New York: World Bank, 2010), 133–143; Geraldine Terry, "No Climate Justice without Gender Justice: An Overview of the Issues," *Gender & Development* 17, no. 1 (2009): 5–18.

19   Sandra G. Harding, "After the Neutrality Ideal: Science, Politics, and 'Strong Objectivity,'" *Social Research* 59, no. 3 (1992): 567–587; Alcoff and Potter, *Feminist Epistemologies*.

20   Gaard, "Ecofeminism Revisited," 26–53.

21   Andrew Jarvis and David Leedal, "The Geoengineering Model Intercomparison Project (GeoMIP): A Control Perspective," *Atmospheric Science Letters* 13, no. 3 (2012): 157; Ben Kravitz et al., "The Geoengineering Model Intercomparison Project Phase 6 (GeoMIP6): Simulation Design and Preliminary Results," *Geoscientific Model Development* 8, no. 10 (2015): 3379–3392; Judith Hauck et al., "Iron Fertilisation and Century-Scale Effects of Open Ocean Dissolution of Olivine in a Simulated CO_2 Removal Experiment," *Environmental Research Letters* 11, no. 2 (2016): 024007.

22   Harding, "Rethinking Standpoint Epistemology," 437–470.

23   Bronson, "Geoengineering," 84.

24   Buck, Gammon, and Preston, "Gender and Geoengineering," 657.

25   Na'ilah Nasir, *Racialized Identities: Race and Achievement among African American Youth* (Palo Alto, Calif.: Stanford University Press, 2011), 5; Ann Louise Keating, "Interrogating 'Whiteness,' (De)Constructing 'Race,'" *College English* 57, no. 8 (1995): 901–918; Michael Omi and Howard Winant, *Racial Formation in the United States* (New York: Routledge, 2014).

26 Liana Christin Landivar, "Disparities in STEM Employment by Sex, Race, and Hispanic Origin," *Education Review* 29, no. 6 (2013): 911–922; US News Staff, "The 2015 U.S. News and Raytheon STEM Index," *US News*, June 29, 2015, http://www.usnews.com/news/stem-index/articles/2015/06/29/the-2015-us-news-raytheon-stem-index; Aaron A. Velasco and Edith Jaurrieta de Velasco, "Striving to Diversify the Geosciences Workforce," *Eos, Transactions American Geophysical Union* 91, no. 33 (2010): 289–290.

27 The Royal Society, *Geoengineering the Climate: Science, Governance and Uncertainty* (London: Royal Society, 2009).

28 Phoebe C. Godfrey, "Introduction: Race, Gender and Class and Climate Change," *Race, Gender & Class* (2012): 3–11; Andrew Baldwin, "Resilience and Race, or Climate Change and the Uninsurable Migrant: Towards an Anthroporacial Reading of 'Race,'" *Resilience* 5, no. 2 (2017): 129–143; Andrew Szasz and Michael Meuser, "Environmental Inequalities: Literature Review and Proposals for New Directions in Research and Theory," *Current Sociology* 45, no. 3 (1997): 99–120.

29 Harding, "Is Science Multicultural?," 301–330; Sandra Harding, "Postcolonial and Feminist Philosophies of Science and Technology: Convergences and Dissonances," *Postcolonial Studies* 12, no. 4 (2009): 401–421.

30 Sandra G. Harding, "Science Is 'Good to Think With,'" *Social Text*, nos. 46–47 (1996): 17.

31 The connection between racism and colonialism/postcolonialism has to do with the ways in which discriminatory laws, rules, discourses, images, and assumptions about racialized groups established under colonialism continue to persist today. These practices shape who gets to speak and whose interests are represented in all spheres of life, including science. This chapter, however, focuses on environmental racism in particular.

32 Davis, "Environmental Oppression"; Peter S. Wenz, "Just Garbage: The Problem of Environmental Racism," in *Environmental Ethics: Readings in Theory and Application*, ed. Louis P. Pojman, Paul Pojman, and Katie McShane (Andover, Mass.: Cengage, 2017), 596–601.

33 Martin V. Melosi, "Equity, Eco-racism and Environmental History," *Environmental History Review* 19, no. 3 (1995): 9.

34 Michael Egan, "Subaltern Environmentalism in the United States: A Historiographic Review," *Environment and History* 8, no. 1 (2002): 26–27; Robert D. Bullard, ed., *Confronting Environmental Racism: Voices from the Grassroots* (New York: South End, 1993), 21; Benjamin A. Goldman, *Not Just Prosperity: Achieving Sustainability with Environmental Justice* (Washington, D.C.: National Wildlife Federation, 1993); Dorceta Taylor, *Toxic Communities: Environmental Racism, Industrial Pollution, and Residential Mobility* (New York: NYU Press, 2014).

35 Ghassan Hage, *Is Racism an Environmental Threat?* (New Jersey: John Wiley & Sons, 2017). Geographers have undertaken some of the most salient work on race and climate change, with more recent studies highlighting sociospatial relations. See, for example, Susannah Fisher, "The Emerging Geographies of Climate Justice," *Geographical Journal* 181, no. 1 (2015): 73–82; and Sam Barrett, "The Necessity of a Multiscalar Analysis of Climate Justice," *Progress in Human Geography* 37, no. 2 (2013): 215–233.

36 Susan J. Smith, "Residential Segregation and the Politics of Racialization," in *Racism, the City and the State*, ed. M. Cross and M. Keith (London: Routledge, 1993), 128–143; Ruth Wilson Gilmore, "Globalisation and US Prison Growth: From Military Keynesianism to Post-Keynesian Militarism," *Race & Class* 40, nos. 2–3 (1999): 171–188; Kay J. Anderson, "The Idea of Chinatown: The Power of Place and Institutional Practice in the Making of a Racial Category," *Annals of the Association of American Geographers* 77, no. 4 (1987): 580–598.

37 Laura Pulido, "Rethinking Environmental Racism: White Privilege and Urban Development in Southern California," *Annals of the Association of American Geographers* 90, no. 1 (2000): 13.

38 Marco Martuzzi, Francesco Mitis, and Francesco Forastiere, "Inequalities, Inequities, Environmental Justice in Waste Management and Health," *European Journal of Public*

Health 20, no. 1 (2010): 21; Florence Lansana Margai, "Health Risks and Environmental Inequity: A Geographical Analysis of Accidental Releases of Hazardous Materials," *Professional Geographer* 53, no. 3 (2001): 422–434; Marianne Lavelle and Marcia Coyle, "Unequal Protection: The Racial Divide in Environmental Law," *National Law Journal* 15, no. 3 (1992): S1–S12.

39 Jeremy Pais, Kyle Crowder, and Liam Downey, "Unequal Trajectories: Racial and Class Differences in Residential Exposure to Industrial Hazard," *Social Forces* 92, no. 3 (2013): 1189–1215; Carl A. Zimring, *Clean and White: A History of Environmental Racism in the United States* (New York: NYU Press, 2017).

40 Philip J. Rasch et al., "An Overview of Geoengineering of Climate Using Stratospheric Sulphate Aerosols," *Philosophical Transactions of the Royal Society A: Mathematical, Physical and Engineering Sciences* 366, no. 1882 (2008): 4007–4037; Douglas G. MacMartin et al., "Geoengineering with Stratospheric Aerosols: What Do We Not Know after a Decade of Research?," *Earth's Future* 4, no. 11 (2016): 543–548.

41 Ingrid Eckerman, "The Bhopal Gas Leak: Analyses of Causes and Consequences by Three Different Models," *Journal of Loss Prevention in the Process Industries* 18, nos. 4–6 (2005): 213–217; B. Bolin and L. C. Kurtz, "Race, Class, Ethnicity, and Disaster Vulnerability," in *Handbook of Disaster Research*, ed. H. Rodriguez et al. (London: Springer, 2018), 181–203; Gabriel Eweje, "Environmental Costs and Responsibilities Resulting from Oil Exploitation in Developing Countries: The Case of the Niger Delta of Nigeria," *Journal of Business Ethics* 69, no. 1 (2006): 27–56; Robert D. Bullard, "Confronting Environmental Racism in the Twenty-First Century," *Global Dialogue* 4, no. 1 (2002): 34.

42 Canadian Press, "Haida Gwaii Iron Dumping Hot Topic as Nations Negotiate International Geoengineering Treaty," *Times Colonist*, October 28, 2012, http://www.timescolonist.com/news/haida-gwaii-iron-dumping-hot-topic-as-nations-negotiate-international-geoengineering-treaty-1.22800; Lukacs, "World's Biggest Geoengineering Experiment," 1–3.

43 See Schneider and Fuhr, this volume, chap. 4.

44 Jeff Tollefson, "Iron-Dumping Ocean Experiment Sparks Controversy," *Nature News* 545, no. 7655 (2017): 393; Kate Elizabeth Gannon and Mike Hulme, "Geoengineering at the 'Edge of the World': Exploring Perceptions of Ocean Fertilisation through the Haida Salmon Restoration Corporation," *Geo: Geography and Environment* 5, no. 1 (2018): e00054.

45 Louise Takeda and Inge Røpke, "Power and Contestation in Collaborative Ecosystem-Based Management: The Case of Haida Gwaii," *Ecological Economics* 70, no. 2 (2010): 178–188; Ken S. Coates, *A Global History of Indigenous Peoples: Struggle and Survival* (London: Palgrave Macmillan, 2004).

46 Gannon and Hulme, "Geoengineering at the 'Edge of the World,'" 4.

47 Buck, Gammon, and Preston, "Gender and Geoengineering."

48 Amy Hinterberger, "Curating Postcolonial Critique," *Social Studies of Science* 43, no. 4 (2013): 619–627.

49 B. F. Chavis Jr., preface to *Unequal Protection: Environmental Justice and Communities of Color*, ed. Robert D. Bullard (San Francisco: Sierra Club Books, 1994), xi–xii.

50 Leila Scannell and Robert Gifford, "Personally Relevant Climate Change: The Role of Place Attachment and Local versus Global Message Framing in Engagement," *Environment and Behavior* 45, no. 1 (2013): 60–85; Jennifer A. Fresque-Baxter and Derek Armitage, "Place Identity and Climate Change Adaptation: A Synthesis and Framework for Understanding," *Wiley Interdisciplinary Reviews: Climate Change* 3, no. 3 (2012): 251–266.

51 Scannell and Gifford, "Personally Relevant Climate Change"; Fresque-Baxter and Armitage "Place Identity and Climate Change Adaptation."

52 Marife M. Ballesteros, "Linking Poverty and the Environment: Evidence from Slums in Philippine Cities" (paper no. 2010-33, PIDS Discussion Paper Series, 2010); Éloi Laurent,

"Issues in Environmental Justice within the European Union," *Ecological Economics* 70, no. 11 (2011): 1846–1853; Michelle L. Bell and Keita Ebisu, "Environmental Inequality in Exposures to Airborne Particulate Matter Components in the United States," *Environmental Health Perspectives* 120, no. 12 (2012): 1699–1704.

53 Joan Martínez-Alier, "Environmental Justice and Economic Degrowth: An Alliance between Two Movements," *Capitalism Nature Socialism* 23, no. 1 (2012): 51–73; John Bellamy Foster, *Ecology against Capitalism* (New York: NYU Press, 2002); Bram Büscher and Murat Arsel, "Introduction: Neoliberal Conservation, Uneven Geographical Development and the Dynamics of Contemporary Capitalism," *Tijdschrift voor economische en sociale geografie* 103, no. 2 (2012): 129–135; Adrian Parr, "The Wrath of Capital: Neoliberalism and Climate Change Politics–Reflections," *Geoforum* 62 (2015): 70–72.

54 Anna Kaijser and Annica Kronsell, "Climate Change through the Lens of Intersectionality," *Environmental Politics* 23, no. 3 (2014): 417–433; Osborne, "Intersectionality and Kyriarchy," 130–151.

55 Tor Håkon Inderberg et al., eds., *Climate Change Adaptation and Development: Transforming Paradigms and Practices* (London: Routledge, 2014).

56 *Sexuality* and *disability* are other significant markers of identity that intersect with *power* and *privilege* in ways that require interrogation as well. This discussion of race, gender, class, and geoengineering is by no means exhaustive.

57 Julia Pongratz et al., "Crop Yields in a Geoengineered Climate," *Nature Climate Change* 2, no. 2 (2012): 101; Andy Jones et al., "The Impact of Abrupt Suspension of Solar Radiation Management (Termination Effect) in Experiment G2 of the Geoengineering Model Intercomparison Project (GeoMIP)," *Journal of Geophysical Research: Atmospheres* 118, no. 17 (2013): 9743–9752; Peter J. Irvine, Andy Ridgwell, and Daniel J. Lunt, "Assessing the Regional Disparities in Geoengineering Impacts," *Geophysical Research Letters* 37, no. 18 (2010).

58 Ben Kravitz et al., "The Geoengineering Model Intercomparison Project (GeoMIP)," *Atmospheric Science Letters* 12, no. 2 (2011): 162–167; Charles L. Curry et al., "A Multimodel Examination of Climate Extremes in an Idealized Geoengineering Experiment," *Journal of Geophysical Research: Atmospheres* 119, no. 7 (2014): 3900–3923. This does not necessarily require models that reflect these concerns initially but, more likely, the integration of these issues into the analysis at the very least. Modeling innovations in population health and bioscience that have incorporated intersectionality into their models could serve as a template. Greta R. Bauer, "Incorporating Intersectionality Theory into Population Health Research Methodology: Challenges and the Potential to Advance Health Equity," *Social Science & Medicine* 110 (2014): 10–17; Ursula A. Kelly, "Integrating Intersectionality and Biomedicine in Health Disparities Research," *Advances in Nursing Science* 32, no. 2 (2009): E42–E56.

59 Hancock, *Intersectionality*.

60 See Jesse L. Reynolds, Andy Parker, and Peter Irvine, "Five Solar Geoengineering Tropes That Have Outstayed Their Welcome," *Earth's Future* 4, no. 12 (2016): 562–568, for a robust critique of this position, which they contend does not adequately account for the ability to calibrate the level of cooling and the chance that increased river flow could offset any decrease in precipitation. However, it is the case that even these more optimistic outcomes assume idealized conditions (i.e., that the level of cooling can be agreed upon and that river flow has not been disrupted by diversion, damming, and climate change).

61 William C. G. Burns, "Climate Geoengineering: Solar Radiation Management and Its Implications for Intergenerational Equity," *Stanford Journal of Law, Science & Policy* 4 (2011): 39–55; Victor Brovkin et al., "Geoengineering Climate by Stratospheric Sulfur Injections: Earth System Vulnerability to Technological Failure," *Climatic Change* 92, nos. 3–4 (2009): 243–259; Albert Lin, "Balancing the Risks: Managing Technology and Dangerous Climate Change," *Issues In Legal Scholarship* 8 (2009): 1–15.

62 Rita Kaur Dhamoon, "Considerations on Mainstreaming Intersectionality," *Political Research Quarterly* 64, no. 1 (2011): 235.

63 "Statistics of Uttar Pradesh," Government of Uttar Pradesh, accessed February 22, 2018, http://up.gov.in/upstateglance.aspx.

64 Aparna Mitra, *Sensitivity of Mangrove Ecosystem to Changing Climate*, vol. 62 (India: Springer, 2013), 143–157; R. Bhatla, Swaleha Tabassum, and A. Tripathi, "Trend Analysis and Extreme Events of Temperature during Post Monsoon and Winter Seasons over Varanasi, Uttar Pradesh, India," *Journal of the Indian Geophysical Union* 20, no. 1 (January 2016): 123–127.

65 Aparna Mitra, "The Status of Women among the Scheduled Tribes in India," *Journal of Socio-economics* 37, no. 3 (2008): 1202–1217; Manoj Pandey and Avaneendra Mishra, "Uttar Pradesh a Brief Analysis," *International Journal of Management, IT and Engineering* 5, no. 3 (2015): 81.

66 Robert Chambers, "Vulnerability, Coping and Policy (Editorial Introduction)," *IDS Bulletin* 2, no. 2 (2006): 1.

67 Naresh Chandra Sourabh, "The Culture of Women's Housework: A Case Study of Bihar, India" (PhD diss., Department of Sociology, University of Helsinki, 2007), http://helda.helsinki.fi/bitstream/handle/10138/23375/thecultu.pdf?sequence=2; Margaret Alston, "Gender Mainstreaming and Climate Change," in *Women's Studies International Forum*, vol. 47 (Oxford: Pergamon, 2014), 287–294; Sirpa Tenhunen, *Secret Freedom in the City: Women's Wage Work and Agency in Calcutta*, vol. 2 (Spotlight Poets, 2003).

68 Md Sadequr Rahman, "Climate Change, Disaster and Gender Vulnerability: A Study on Two Divisions of Bangladesh," *American Journal of Human Ecology* 2, no. 2 (2013): 72–82; Sandra Wachholz, "'At Risk': Climate Change and Its Bearing on Women's Vulnerability to Male Violence," in *Issues in Green Criminology*, ed. Piers Beirne and Nigel South (London: Willan, 2013), 183–207. Internal displacement in India today tends to be a result of politicized development, secessionist movements, communal conflict, local violence, and natural disasters. India has no IDP policy and has not signed the U.N. Refugee Convention or its accompanying protocol. Mahendra P. Lama, "Internal Displacement in India: Causes, Protection and Dilemmas," *Forced Migration Review* 8 (2000): 5.

69 Roger Bennett and Rita Kottasz, "Advertising Imagery Employed by Disaster Relief Organisations and Media Stereotyping of the Recipients of Aid," in *New Meanings for Marketing in a New Millennium* (London: Springer, 2015), 225–229.

70 Susan D. Moeller, *Compassion Fatigue: How the Media Sell Disease, Famine, War and Death* (New York: Routledge, 2002); Johannes von Engelhardt and Jeroen Jansz, "Distant Suffering and the Mediation of Humanitarian Disaster," in *World Suffering and Quality of Life*, ed. Ronald E. Anderson (Dordrecht, Netherlands: Springer, 2015), 75–87.

71 Úrsula Oswald Spring, "Social Vulnerability, Discrimination, and Resilience-Building in Disaster Risk Reduction," in *Coping with Global Environmental Change, Disasters and Security*, ed. Hans Günter Brauch et al. (Berlin: Springer, 2011), 1169–1188.

72 Jan Culik, "Anti-immigrant Walls and Racist Tweets: The Refugee Crisis in Central Europe," The Conversation, July 24, 2015, http://theconversation.com/anti-immigrant-walls-and-racist-tweets-the-refugee-crisis-in-central-europe-43665; Stephen Zunes, "Europe's Refugee Crisis, Terrorism, and Islamophobia," *Peace Review* 29, no. 1 (2017): 1–6.

73 Singh, "Issues of Farmers," 89.

74 O'Brien, "'First Be Reconciled.'"

75 Daniel P. Aldrich, "The Externalities of Strong Social Capital: Post-tsunami Recovery in Southeast India," *Journal of Civil Society* 7, no. 1 (2011): 81–99.

76 Amarnath Tripathi, "Socioeconomic Backwardness and Vulnerability to Climate Change: Evidence from Uttar Pradesh State in India," *Journal of Environmental Planning and Management* 60, no. 2 (2017): 328–350; Andrew G. Turner and Hariharasubramanian

Annamalai, "Climate Change and the South Asian Summer Monsoon," *Nature Climate Change* 2, no. 8 (2012): 587.

77 Pyaralal Raghavan, "Income of Uttar Pradesh Farmers Are among the Lowest in the Country," *Times of India*, April 3, 2017, http://blogs.timesofindia.indiatimes.com/minorityview/income-of-uttar-pradesh-farmers-are-among-the-lowest-in-the-country/; Ajit Kumar Singh, "Income and Livelihood Issues of Farmers: A Field Study in Uttar Pradesh," *Agricultural Economics Research Review* 26 (2013): 89. These numbers are from 2013, which is the last time comprehensive statistics were available (according to the 2013 exchange rate, this would amount to approximately $90); the symbol $ refers to U.S. dollars.

78 Jonathan Kennedy and Lawrence King, "The Political Economy of Farmers' Suicides in India: Indebted Cash-Crop Farmers with Marginal Landholdings Explain State-Level Variation in Suicide Rates," *Globalization and Health* 10, no. 1 (2014): 16; Anoop Sadanandan, "Political Economy of Suicide: Financial Reforms, Credit Crunches and Farmer Suicides in India," *Journal of Developing Areas* (2014): 287–307.

79 Amita Shah and Baidyanath Guru, *Poverty in Remote Rural Areas in India: A Review of Evidence and Issues* (Ahmedabad, India: Gujarat Institute of Development Research, 2003); Nancy Kaushik and S. C. Rai, "Economic Structure of Drought Prone Regions of India," *Researchers World* 8, no. 3 (2017): 84.

80 P. Venkata Rao and Hari Charan Behera, "Agrarian Questions under Neoliberal Economic Policies in India: A Review and Analysis of Dispossession and Depeasantisation," *Oriental Anthropologist* 17, no. 1 (2017): 17; Anthony P. D'Costa, "Land, Livelihoods, and Late Capitalist Development," in *The Land Question in India: State, Dispossession, and Capitalist Transition*, ed. Anthony P. D'Costa and Achin Chakraborty (Oxford: Oxford University Press, 2017), 325–330.

Part III

State Power, Economic Planning, and Geoengineering

8

Mobilizing in a Climate Shock

Geoengineering or Accelerated
Energy Transition?

LAURENCE L. DELINA

"Nature Roars. Washington Hears Nothing." This is the headline of an editorial penned by the editorial board of the *New York Times* on September 16, 2018.[1] The piece itemizes heat waves, droughts, megafires, and hurricanes—all extreme events that occurred during the summer of 2018. As the piece was published, Hurricane Florence had just drenched the Eastern Seaboard of the United States while Typhoon Mangkhut had just pummeled the Philippines and was on its way to Hong Kong. Also in the summer months, wildfires raged in California, Greece, and as far north as the Arctic Circle. Japan recorded its highest temperature in history in a heat wave that killed sixty-five people in a week. Europe also experienced scorching temperatures. As the climate continues to warm, extreme weather events we witnessed during the summer of 2018 are projected to increase in numbers and intensity. But the linkage was ignored specifically in President Trump's Washington.

With high certainty, scientific knowledge leaves no doubt that anthropogenic sources are the cause of the increasing concentration of atmospheric greenhouse gases (GHGs) that trigger the ongoing changes in the earth's ecological systems. With the devastating impacts of extreme weather events that occurred at the current warming of 1°C, we now need to prepare for the likely increase beyond 2°C within the next thirty years. Left unabated, our current and future emissions, however, will most likely lead us into a warming of more than 6°C by the end of this century, a line that once crossed jeopardizes life on earth.[2] Such warming means "substantial species extinction, global and regional food insecurity, consequential constraints on common human activities and limited potential for adaptation in some cases."[3] These so-called fat-tail risks, however, remain one of the most underappreciated aspects in the climate change challenge.

In terms of probabilities, climate change displays not a "normal" distribution but a "heavy-tailed" or "fat-tailed" distribution. Unlike the shape of a bell in the normal distribution curve, there is more area under the far right extreme of the curve, meaning a greater likelihood of warming that is well in excess of the *average* amount of warming predicted by climate models. Gernot Wagner and Martin Weitzman, in their book *Climate Shock*, show that the likelihood of exceeding 6°C warming is not 2 percent (as it would be in a normal distribution) but close to 10 percent (it is therefore fat tailed).[4] A warming to this extent increases the likelihood that we would exceed certain "tipping points," such as the rapid melting of large parts of Greenland and the Antarctic ice sheet, hence producing massive sea level rise, which could impact hundreds of thousands, if not millions, of people living in low-lying islands and coastal regions. Melting, however, could occur faster than the models have indicated. Thus the fat tail should not be neglected but rather appreciated as an argument for acting swiftly. So *what if* one day we reach climate tipping points and right before our eyes are the climate shocks of our lifetime? Let us imagine what we could do.

An (Abrupt) Emotional Response? A Crisis Stance

Acting on public clamor to act now and swiftly to avoid further future damages, governments *would* enter a "state of exception," suspending the normal course of events and instead adopting extraordinary measures.[5] Under this condition, states elect to inject stratospheric aerosol to reduce incoming solar radiation and thus induce an artificial cooling effect. This engineering technology, called solar radiation management (SRM), is foremost among the many geoengineering techniques that in recent years have started to gain popularity in scientific circles (e.g., Royal Society[6]) and even received funding support from the likes of Bill Gates. SRM, according to one of its advocates, David Keith, is "cheap and technically easy."[7] Keith's SRM proposal involves simulating the eruption of the Pinatubo volcano—a natural event in the 1990s that led to global cooling—using a fleet of twenty Gulfstream G650 jets that would inject 25,000 tons of sulfuric acid vapor into the stratosphere in the first year, then 50,000 in the next, then 250,000 after a decade, and then 1,000,000 after a half century. Keith's "artificial volcanoes," according to his estimates, will cost less than $6 billion[8] for a decade—way below the $300 billion annual spending on sustainable energy technologies. Keith's simulations show that with SRM, the earth would still warm by 1°C. For SRM to be effectively deployed at a planetary scale, however, Keith suggests that governments must be willing to tolerate some central controls to facilitate its regulation and optimization.

SRM, nonetheless, remains an unproven technology, and its risks have been thoroughly described (see the introduction chapter of this volume). In short, SRM could affect monsoons and precipitation patterns at the same time that it could pollute the air. SRM is a process that involves an unending masking of the atmosphere to redirect solar energy. With rising GHG emissions, those "artificial volcanoes" need to "erupt" perpetually. *What if* something goes wrong?

What if some parts of the artificial volcanoes break down? *What if* its dreaded effects on precipitation patterns result in affected states taking drastic measures of their own to try to counter the negative impacts?

Termination of SRM schemes would lead to a termination shock.[9] When abruptly terminated, Matthews and Caldeira warn that current rates of warming could increase by a factor of ten to twenty.[10] Such termination carries substantial risk of very high rates of warming.[11] Instead of curtailing warming, failed SRM could easily trigger a spiral of climate shocks, contravening the very reason it was put up in the first place. Even in a case where geoengineering successfully arrests further warming, the technological shelf life of these technologies far exceeds the intent of its engineers, opening up challenges as to how its long-term, most likely intergenerational (if not perpetual) consequences should and could be monitored and managed. Once deployed, these technologies may no longer be malleable and flexible. We will be locked into the consequences, which remain unknown. Absent a governance framework for these risky technologies,[12] causality and liability will be impossible to attribute should these many negative consequences occur.[13] When that happens, international conflicts and strains in international relations are most likely,[14] hence exacerbating planetary chaos.

Prospects and support for geoengineering technologies are not mainly triggered by honest visions of a climate-safe and livable planet. The mirage of ease, accessibility, and affordability that SRM advocates are heralding has to be noted in conjunction with the actors and interests behind it. Andreas Malm remarks, "No special intelligence is needed to decode the interests at work here, . . ." after naming a Canadian oil magnate as among the generous funders of the study of geoengineering and think tanks in the conservative right as its key promoters.[15] (Although contributions to this volume by Ina Möller, chap. 2, and by Lili Fuhr and Linda Schneider, chap. 4, suggest this is no longer the case, as more liberal actors have started funding and promoting geoengineering.)

Geoengineering is an option that does not necessarily require putting a brake on the profitable business of emission spewing; instead, it floats the idea that climate change can be contained without a complete shift away from fossil fuels. Geoengineering also cements the idea promoted by neoliberalism—that growth without end can continue even at the expense of people's lives and their livelihoods and the natural environment. Geoengineering fuels climate denialism, where politicians persist in playing the ostrich in full denial mode. At a time when we should be focusing our resources and capacities on reducing current and future emissions using less-risky, environmentally benign and perpetually available wind, water, and sunlight energy, geoengineering represents a costly diversion.

When fat-tailed climate shocks come to pass, *what if* governments, instead of overly relying on geoengineering techniques such as SRM, take a more reasoned stance—that is, to address the climate change challenge at the core of what has been driving it: Anthropogenic GHG emissions? *What if* governments embark on a cooperative coalition to undertake ultraradical emission cuts instead?

A More Reasoned Response? A Stance of Planetary Stewardship

So far, the only demonstrated and verified option for substantial emissions reduction is to focus on the energy sector—which holds the lion's share in anthropogenic GHG emissions. This option entails a rapid, systematic, and structured retirement of fossil-based electricity generation and replacement with 100 percent renewable energy technologies complemented by large-scale energy efficiency and conservation programs that result in avoiding electricity consumption whenever possible. Study after study on the possibility of gargantuan transitions of this scale has been churned out in the last decade. Notable among them is one by Stanford University scholars Mark Jacobson and his colleagues, who showed simulations on a country-by-country basis of what it would take for the global energy system to transition toward 100 percent wind, water, and sunlight energy technologies.[16] But they are not alone; many others have also written about this technological possibility.[17] Almost in chorus, these studies suggest that a renewable energy–powered world can be accomplished using commercially available sustainable energy technologies but, at the same time, issue caveats about its limitations. The present technological challenges, according to these studies, now rest on intensifying research and development of key yet immature infrastructure, such as smart grids and storage technologies.

The concept of a sustainable energy transition is not entirely new. During the energy crises of the 1970s and 1980s, this idea had centrally occupied discussions as the possible solution. As early as 1981, NASA's James Hansen and colleagues had called for energy transition as a climate change response, writing in *Science* that "[an] appropriate strategy may be to encourage energy conservation and develop alternative energy sources."[18] Sustainable energy refers to the form of energy obtained from renewable resources emitting the lowest GHGs in its life cycle. These are energy from wind, water, and sunlight. The technologies used to harness these sources (wind turbines, hydropower, and solar photovoltaics) and convert them to energy services (electricity, transport, heating, and cooling) are collectively called sustainable energy technologies, which also include technologies for energy conservation and energy efficiency in utilities, buildings, appliances, and industrial equipment.

And while these technologies are already available commercially with their costs quickly declining, the speed by which they are deployed remains slow. The reasons for this are many, but in the context of rapid climate action, free riding has been blamed, mainly due to the absence of incentives to accelerate energy transition.[19] The free market orientation, where economic growth at the expense of the environment is the core objective, also precludes many governments from adopting stringent policies to facilitate rapid energy transition.[20] Big businesses from high-emission sectors are also largely responsible for the transition's slow speed.[21]

Bringing the transition up to speed would therefore require a radical change not only in technological systems but also in governance to marshal and oversee the rapid deployment of commercially available sustainable energy technologies.[22] Since these activities need financial and labor investments on a scale beyond the ordinary, this massive turn would entail a return to planning with the state acting at an extraordinary time. A shift to planning, fortunately, is not entirely new—even in capitalist governments.

The intensification of Allied mobilization during the Second World War offers one lens to view such extraordinariness. The mobilization in the United States, which amounted to more than $4 trillion in today's money, helped ensure that Europe and the Pacific were not lost to the Axis powers. Nobody now would suggest that the decision by the United States to spend such huge sums was wrong. The American economy, as wartime mobilization demonstrated, can be turned into an engine of production to counter a common threat. Wartime history shows that munitions can be quickly produced and deployed to wherever most needed, that public funding was essential to accomplish these feats, and that labor power can be quickly mobilized. The state, this history reveals, had the capacity to reorganize its administrative agents and institutions to direct resources, requisition properties, sell bonds, terminate the production of "competing" goods—private cars and other luxuries in this case—and summon essential labor power.[23] In the same vein, a rapid climate mitigation program where sustainable energy transitions are quickly accomplished *could* be patterned after this history.[24]

To effectively and quickly transition to 100 percent sustainable energy systems—beyond electricity services to also include sustainable energy use in transport and industries, including shipping and aviation—a similar centralization of power at an extraordinary moment in history is required. Mimicking the mobilization during the Second World War, a special *Ministry for Rapid Mitigation* will be afforded administrative power a notch higher than conventional ministries to plan and coordinate the transition.[25] The powers of this ministry, which will operate under a "state of exception" (i.e., beyond the established rules of the game and as a special kind of political organization),[26] would include organizing the mass production of wind, water, and sunlight technologies and systems and deploying them quickly; raising large sums of money; restricting luxurious consumption; redirecting labor; and accelerating the research and development of storage technologies, among other things. The entire venture would take approximately fifteen years for historical top twenty high-emission states, twenty years for other industrialized countries, twenty-five for developing countries in the top twenty emitters, and up to forty years for the rest.[27] The analogy, nonetheless, has its challenges in that it seems "unlikely, that governments will adopt the Project without strong countervailing pressure for action from the public, robust social, political, technical and economic incentives to undertake these activities, a legally binding international climate agreement, or a combination of the above."[28]

Weighing Our Options

Given that our time to effectively respond to the climate challenge is now too short and the institutional obstacles are firmly secured by very powerful interests and/or myopic governments, business as usual and market-oriented responses are less likely to produce the necessary shifts to ensure a climate-safe environment for all. We need to approach our common challenge by rethinking the sociopolitical-economic paradigms that have long defined our contemporary governance structures—without waiting for climate fat tails to eventuate.

Climate scientists Kevin Anderson and Alice Bows stress the need to address issues "about the structure, values and framing of contemporary society."[29] Anderson, in reaction to the release of the fifth assessment report produced by the Intergovernmental Panel on Climate Change (IPCC) in 2013—which proposed that putting a price on emissions would force producers and consumers toward energy transition—said, "I hold that such a market-based approach is doomed to failure and is a dangerous distraction from a comprehensive regulatory and standard-based framework."[30] In a 2017 interview with the *Nation*, Anderson stated, "The dominant neoclassical, market-based approach I think is fundamentally flawed for guessing system level changes. It is arguably very good for looking at small changes within certain productive parts of society: sock or shoe manufacturers, possibly even motorbike manufacturers. But once it gets to planes or cars or climate change, it has nothing useful to say. It's a field that has overplayed its hand and been accepted by all of us as important, and has helped undermine any real mitigation for a long time."[31]

For those who worship in the temple of capitalism, the idea of a centrally planned economy is revolting. After all, was not victory in the Cold War a manifestation of the triumph of the markets and the defeat of planning? In a climate emergency, geoengineering would be more seductive than a special Ministry for Rapid Mitigation simply because the former resonates well with the ideals of capitalism—it is a magical pill securing the status quo; there is no need for emissions reduction at all—whereas the latter would result in the end of what Malm[32] calls fossil capital, a call for total makeover of our sociopolitical-economic order. It is highly likely, therefore, that in an event of a climate shock, when the impacts of climate change could no longer be questioned, myopic governments such as President Trump's would put their weight behind the geoengineering option.[33]

Geoengineering—while parading its support for free market ideals—is, when appreciated better, actually about perpetuating a top-down, centralized-only state system, since it requires a central controller to attend to the thermostat.[34] Sustainable energy transitions, by contrast and despite their nod to planning, open up for opportunities for greater public participation, since these processes could occur at multiple levels and could be acted upon by multiple actors, not just a central planner. Energy transitions allow opportunities for a polycentric approach to governance where every pocket and space of transition can be vested the power to facilitate, administer, manage, and operate their own sociotechnical energy systems—in short, govern them.[35] The energy transition envisaged from a wartime mobilization model can thus be cascaded across several levels of governance, from the international to the regional to the national and to the local, unlike geoengineering, which necessitates an international-only kind of governance. Such an opening up to social participation and public engagement contests the overly Promethean approach in the deployment of geoengineering technologies. Importantly, the morality of the challenge of climate change is also brought to bear when choosing which technological approach would best fit at a time of climate emergency. Geoengineering, on the one hand, frees us from responsibility for the behavior that creates the problem of climate change in the first place. Energy transition, which addresses climate change at its source, on the other hand, faces the cause of the problem front and center.

Climate change action is a multifaceted affair of organized collective responsibility to protect ecosystems and the services they provide us. But it is only possible to respond to crises when we identify the danger. The danger of climate shocks, this chapter suggests, is one way to *imagine* these potential crises. Through this gedankenexperiment, this "thought experiment," we have imagined how a catastrophic cascade of extreme climate shocks could help us explore our potential future governance responses. This chapter has explored two futures. Sustainable energy systems transitions, on the one hand, address the causes of climate change and could lead to a safe and sustainable global energy supply system that can exist forever. On the other hand, geoengineering cannot achieve that; instead, its fatal consequences, despite its promise to reverse warming, mean that these options are risky Band-Aid approaches. When limited resources—time most especially—must be efficiently spent, geoengineering only reduces and crowds out our available time, which can be better spent toward the diffusion of already proven and demonstrated sustainable energy technologies.

We are clearly short on time. Climate science has already provided us with some indicators of when emission reductions should have started. The IPCC,[36] in its fifth assessment report, suggested emissions must peak before 2030, decline sharply thereafter, reach zero by around 2070, and register negative values by 2100. The German Advisory Council on Global Change[37] suggested an early peak by 2020 and a fall to zero levels by 2040. Anderson and Bows[38] are more stringent, noting that emissions must peak from 2016 to 2022 and be followed by rapid reductions to near zero by 2050. If delayed, Anderson and Bows warn that the maximum reduction rate *must* exceed 9 percent per year and that incremental adjustments to market incentives alone would no longer be sufficient. The IPCC[39] also notes that delay "will require substantially higher rates of emissions reductions from 2030 to 2050: a much more rapid scale-up of low-carbon energy over this period . . . and higher transitional and long-term economic impacts."

We are clearly in a critical decade. Our emissions must be deliberately reduced now to avoid catastrophic warming. We need not wait for climate shocks to occur to start acting. We even do not know whether it is still even possible to clean up that mess. Moving fast enough on this grand scale *now* will require quick and agile decision-making. Unfortunately, that is currently out of reach in many contemporary neoliberal-oriented democratic governments. It seems after all that going to war is easier than facilitating real change for the climate. That should not be the case.

Notes

1 I thank Cynthia Barakatt for her comments and Andreas Malm for the invitation to contribute and for his thoughtful suggestions.

2 James Hansen et al., "Assessing 'Dangerous Climate Change': Required Reduction of Carbon Emissions to Protect Young People, Future Generations and Nature," *PLoS One* 8 (2013): e81648; James Hansen et al., "Target Atmospheric CO_2: Where Should Humanity Aim?," *Open Atmospheric Science Journal* 2, no. 1 (2008): 217–231.

3 International Panel on Climate Change (IPCC), *Climate Change 2014: Synthesis Report of the Fifth Assessment Report of the IPCC* (Geneva, Switzerland: IPCC, 2014), 18.

4 Gernot Wagner and Martin L. Weitzman, *Climate Shock: The Economic Consequences of a Hotter Planet* (Princeton, N.J.: Princeton University Press, 2015).

5 Giorgio Agamben, *State of Exception* (Chicago: University of Chicago Press, 2005); cf. Markus Lederer and Judith Kreuter, "Organising the Unthinkable in Times of Crises: Will Climate Engineering Become the Weapon of Last Resort in the Anthropocene?," *Organization* 25 (2018): 472–490.

6 The Royal Society, *Geoengineering the Climate: Science, Governance and Uncertainty* (London: Royal Society, 2009).

7 David Keith, *A Case for Climate Engineering* (Cambridge, Mass.: MIT Press, 2013), ix.

8 The symbol $ refers to U.S. dollars.

9 Royal Society, *Geoengineering the Climate*.

10 H. Damon Matthews and Ken Caldeira, "Transient Climate-Carbon Simulations of Planetary Geoengineering," *Proceedings of the National Academies of Science* 104 (2007): 9949–9954.

11 Andrew Ross and H. Damon Matthews, "Climate Engineering and the Risk of Rapid Climate Change," *Environmental Research Letters* 4 (2009): 045103.

12 Gareth Davies, "Geoengineering: A Critique," *Climate Law* 1 (2010): 429–441.

13 Mike Hulme, "Climate Intervention Schemes Could Be Undone by Geopolitics," *Yale Environment 360*, June 7, 2010, http://e360.yale.edu/features/climate_intervention _schemes_could_be_undone_by_geopolitics; Mike Hulme, Saffron J. O'Neill, and Suraje Dessai, "Is Weather Event Attribution Necessary for Adaptation Funding?," *Science* 334 (2011): 764–765.

14 Bronislaw Szerszynski et al., "Why Solar Radiation Management Geoengineering and Democracy Won't Mix," *Environment and Planning A* 45 (2013): 2809–2816.

15 Andreas Malm, "Socialism or Barbecue, War Communism or Geoengineering: Some Thoughts on Choices in a Time of Emergency," in *The Politics of Ecosocialism: Transforming Welfare*, ed. Kajsa Borgnäs et al. (Basingstoke, England: Routledge, 2015), 180–194; see also Schneider and Fuhr, this volume, chap. 4.

16 Mark Jacobson et al., "100% Clean and Renewable Wind, Water, and Sunlight All-Sector Energy Roadmaps for 139 Countries of the World," *Joule* 1 (2017): 108–121.

17 For a list and description, see Laurence Delina, *Strategies for Rapid Climate Mitigation: War Mobilisation as Model for Action?* (Abingdon, Va.: Routledge, 2016), 39–43.

18 James Hansen et al., "Climate Impact of Increasing Atmospheric Carbon Dioxide," *Science* 213 (1981): 957–966.

19 William D. Nordhaus, "Climate Clubs: Overcoming Free-Riding in International Climate Policy," *American Economic Review* 105 (2015): 1339–1970.

20 Cf. Naomi Klein, *This Changes Everything: Capitalism vs. the Climate* (New York: Simon & Schuster, 2014).

21 Mark Diesendorf, *Climate Action: A Campaign Manual for Greenhouse Solutions* (Sydney: University of New South Wales Press, 2009).

22 Erik Laes, Leen Gorissen, and Frank Nevens, "A Comparison of Energy Transition Governance in Germany, the Netherlands and the United Kingdom," *Sustainability* 6, no. 3 (2014): 1129–1152.

23 Delina, *Strategies*.

24 For more detailed descriptions of the Second World War climate analogy and its caveats, see Delina; for an abbreviated version, see Laurence Delina and Mark Diesendorf, "Is Wartime Mobilisation a Suitable Policy Model for Rapid National Climate Mitigation?," *Energy Policy* 58 (2013): 371–380. Notably, Lederer and Kreuter have also used a wartime analogy, particularly the early development stages of nuclear weapons, that suggests geoengineering technologies could be organized not necessarily around legalistic means but following the logic of technical feasibility, political acceptance, and bureaucratic momentum. Lederer and Kreuter, "Organising the Unthinkable."

25 Delina, *Strategies*, 148–149.

26 Agamben, *State of Exception*.

27 Delina, *Strategies*, 68–69.

28 Delina, 161.

29 Kevin Anderson and Alice Bows, "Beyond 'Dangerous' Climate Change: Emission Scenarios for a New World," *Philosophical Transactions of the Royal Society A* 369 (2011): 20–44.

30 Tom Bawden, "Plan to Use Financial Markets to Halt Climate Change Is 'Doomed,'" *Independent*, October 1, 2013.

31 Kate Aronoff, "Could a Marshall Plan for the Planet Tackle the Climate Crisis?," *Nation*, November 16, 2017.

32 Andreas Malm, *Fossil Capital: The Rise of Steam Power and the Roots of Global Warming* (London: Verso, 2016).

33 Lederer and Kreuter, "Organising the Unthinkable," 472–490.

34 See Malm, this volume, chap. 10.

35 Laurence Delina and Anthony Janetos, "Cosmopolitan, Dynamic and Contested Energy Futures: Navigating the Pluralities and Polarities of the Energy Systems of Tomorrow," *Energy Research & Social Science* 35 (2018): 1–10.

36 IPCC, *Climate Change 2014*.

37 German Advisory Council on Global Change, *Solving the Climate Dilemma: The Budget Approach* (Berlin: WBGU, 2009).

38 Anderson and Bows, "Beyond 'Dangerous' Climate Change," 20–44.

39 IPCC, *Climate Change 2014*, 25.

9

A Left Defense of Carbon Dioxide Removal

●

The State Must Be Forced to Deploy Civilization-Saving Technology

CHRISTIAN PARENTI

> Experience has shown that only princes
> and republics that have their own army
> make great progress, while mercenary
> armies do nothing but harm.
> —Niccolo Machiavelli, *The Prince*

If civilization was serious about saving itself, powerful and wealthy states would treat the climate crisis like a massive military emergency and do the following: euthanize the fossil fuel industry while rapidly building out a clean energy infrastructure so as to eliminate greenhouse gas (GHG) emissions[1]; and, more controversially, nationalize existing technologies for carbon dioxide removal (CDR) and immediately commence massive crash programs of publicly funded atmospheric CO_2 drawdown.[2] The state must deploy this technology as a public good and manage atmospheric CO_2 levels as a global commons because the costs of CDR are too high and its benefits too broadly distributed for private profit-seeking investment in the technology to make sense. Here the mercenary logic of the market fails.

The science on climate tipping points is clear: even if we stopped all GHG pollution, we would need to strip CO_2 from the atmosphere. At the time of writing, CO_2 concentrations are 405 ppm (parts per million) and need to be at 350 ppm or lower to avoid self-compounding climate breakdown. In other words, stopping CO_2 emissions is not enough; we also need a global program of negative emissions.

In the following paragraphs, I argue that attempts to integrate CDR technology into competitive markets will fail; only deployment by governments, as something like a public utility, can bring CDR to scale. Orthodox economists argue that the proper response to climate change is to simply wait for the market to innovate. Alas, the real history of capitalism does not comport with the laissez-faire mythology. Innovation is not an apolitical, "merely" technical process; rather, it is a highly political process. The technological outcomes of innovation are the products of political struggle just as much as they are products of purely scientific choices.

The good news is that all the technologies necessary to achieve negative emissions already exist in proven form. And states have the power and money to undertake such a task at the necessary scale. However, we lack policies that can bring about the necessary radical technological transformations. In this, we face what Marx described as a contradiction between the forces of production and the relations of production. The social relations of capital are now holding back the full potential of some of the most promising technologies that modern science has yet invented.

We lack the necessary policies because of two ideological problems, or blockages. The first blockage is the market fundamentalism that corrupts intellectual life in the West. This even penetrates much of the left, creating a statephobia among people who, while not deluded about markets and capitalism, still cannot see the usefulness of government power. The second blockage is the deep, often unexamined, technophobia and nature fetish of many environmentalists.

As a result of these ideological blockages, most of the people who support CDR technology operate with ridiculous ideas about market-based mechanisms for the technology's mass deployment. Meanwhile, those on the left and in environmental movements who think critically about capitalism and pressure the government to craft progressive policy remain largely silent on CDR or oppose it as just sophisticated technogreenwashing.[3] Because CDR means large-scale technological intervention into the climate system, most greens reject it without further consideration. This is highly dangerous and wrongheaded.

Ultimately, the challenge of climate change requires that the human species overcome the central social divisions produced by class society and become a fully self-conscious species that can recognize its collective interdependence and embrace our role as environment makers. The existence of powerful new technologies, like CDR, that if rationally deployed could help maximize the chances of human survival should give us hope that we can rise to the challenge.

How CDR Works

An important distinction must be clarified at the outset. So-called carbon engineering or carbon capture and storage (CCS), or more simply CDR, which seeks to mechanically remove CO_2 from wastes streams and the ambient air, should not be conflated with geoengineering in the form of solar radiation management (SRM), which would attempt to artificially increase the earth's albedo, or reflectivity, by spraying reflective particles into the stratosphere in an effort to deflect infrared radiation back out into space before it heats the planet's surface.

Most SRM schemes contemplate spraying mass quantities of sulfate particles into the stratosphere so as to mimic the cooling effects of volcanic eruptions. The problem with this style of SRM is that the sulfate particles would likely wash out of the atmosphere as acid rain, and the SRM-induced cooling could have catastrophic impacts on precipitation patterns.[4] Nor is all CDR equal. For example, so-called ocean iron fertilization (OIF) involves adding iron dust to the sea so as to trigger carbon dioxide–sequestering algal blooms. But overabundant algae can suck up too much oxygen and cause eutrophication. The infamous "dead zone" in the Gulf of Mexico is the by-product of massive fertilizer runoff–fed algal blooms.[5]

On the other hand, mechanical CDR, or direct air capture (DAC), poses far fewer risks but requires far more investment. Instead of adding ingredients to natural systems, DAC simply removes CO_2 from the air. This sort of CDR removing CO_2 from the ambient air is not too complicated. It has been happening in submarines and spacecraft like the U.S. space shuttles for decades. In recent years, Klaus Lackner, director of the Center for Negative Carbon Emissions at Arizona State University, has developed a device similar to an artificial tree that captures CO_2 a thousand times more effectively than actual trees. The device involves strips of plastic coated with a commercially available "anionic exchange" resin; CO_2 in the air binds to the resin and is then washed off with water. Next, the CO_2 is stripped out of the water and stored as pure gas.

The problem with this approach becomes where and how to store CO_2 gas. One solution has involved pressurizing CO_2 into a liquefied form and then injecting it into underground cavities such as depleted oil and gas wells or deep saline aquifers. But such liquefied gas is buoyant and can migrate through cracks to the surface. Being invisible and odorless, leaks of pure carbon dioxide, if trapped in buildings or basements, could easily kill unsuspecting people who wander into it. A greater danger is that vast pools of stored CO_2 could escape back into the atmosphere, defeating the whole purpose of its storage. In short, underground storage of CO_2 gas remains problematic.[6]

However, in 2016, an experiment in Iceland mixed carbon dioxide and hydrogen sulfide into water and then pumped the mixture into underground basalt rock formations. Within two years, the CO_2 in the water mixture "precipitated" into a white chalky solid: a carbonate rock similar to limestone. As one report explained, "The researchers were amazed by how fast all the gas turned into a solid—just two years, compared to the hundreds or thousands of years that had been predicted." In fact, the mineralization happens so quickly that if injected too rapidly, pathways leading deeper into the basalt rocks can clog up before the dissolved CO_2 and water mixture has percolated to its intended depths. Thus along with science and engineering, the injection process (much like oil drilling) involves a bit of art and finesse.[7] Luckily, the process, which uses huge amounts of water, functions perfectly well using salt water; also, no external heat source is required, and basalt, the host rock, is one of the most common types on earth.

Though at one level miraculous, the transformation is just a radically accelerated version of a naturally occurring process. When CO_2 mixes with water, it becomes carbonic acid (H_2CO_3), and when it falls back to earth, the carbonic acid interacts and binds with minerals in the surfaces of rocks, primarily calcium and magnesium, to form various types of calcium carbonate rock such as ankerite, all of which are similar to limestone.

In other words, some rocks actually grow by the slow accretion of calcium carbonate rock produced by mineralization of what was once atmospheric carbon dioxide. In Oman, where carbon-capturing peridotite rock proliferates, "white carbonate minerals run through slabs of dark rock like fat marbling a steak . . . Even pooled spring water that has bubbled up through the rocks reacts with CO_2 to produce an ice-like crust of carbonate that, if broken, re-forms within days."[8] The idea of harnessing and artificially accelerating this process was only first proposed in 1995 by Lackner and a team of researchers.[9]

The current Iceland project operates as part of a large geothermal power plant—this is helpful because drilling deep into rocks and managing the flow and pressure of gases and liquids below ground is a central geothermal skill set. Called CarbFix, the project currently strips and stores five thousand metric tons of CO_2 a year. That only equals the annual emissions of about two thousand cars. Another test project, conducted by the U.S. Department of Energy's Pacific Northwest National Laboratory (PNNL), found that a mixture of mostly liquid CO_2 with only a small amount of water also mineralized within two years.[10] By some estimates, there are seventeen such CDR and mineral storage experiments around the globe.

The point is, human civilization has the technical ability to strip atmospheric CO_2 and safely store it. The stripping of ambient CO_2 and its mineral storage is not a possibility; it is an actuality, a proven technology, operating right now as you read these words.

The problem is how to bring CDR and mineral storage to scale? Here the limits of the first blockage—market thinking—immediately come it to focus. For example, a partner in CarbFix is a Swiss company that, although born of Swiss government funding and spun off from a large public university, now tries to operate as a profit-making venture. Its only real commercial clients seem to be a few greenhouses that use CO_2 gas to enhance plant growth and some beverage companies that use the gas in carbonation and dry ice.[11] Another carbon capture project in Canada plans to sell its by-product, magnesium carbonate, to wastewater treatment facilities and to the steel industry. Another company uses captured CO_2 in foam mattresses. In Australia, free market boosters of enhanced weathering technology push the idea of selling artificially created limestone as building material. The economics of that are patently ridiculous: Why buy expensive rock when cheaper natural rock is available?[12]

One academic paper described the quandary as follows: "At present, there is a large gap between the total cost per ton of CO_2 handled by CCS and the revenue available to operators for capturing and storing CO_2 (for example, from the price of emissions allowances in the EU ETS). For CCS to be attractive, this gap must be closed, through both a higher allowance price induced by a stricter climate policy and technological advances that lower CCS cost."[13]

The market fails because there is no way the world can "use" all potentially captured CO_2. Thus captured CO_2 cannot be disposed of by means of commodification and sale into competitive markets.

The world economy is producing about forty billion metric tons of carbon emissions a year. At current prices—which means little, since there are no real functioning carbon markets and price estimations vary widely—my calculations indicate stripping out that amount of CO_2 could cost up to $24 trillion[14], a sum equal to 133 percent of the annual U.S. GDP. But to be fair, the price of CDR technology and mineral storage is dropping. The cost of all prototype technologies is astronomically high but tends to fall precipitously as production improves and costs decline. Thus let us assume that stripping and storing a year's CO_2 emissions will soon cost only $12 trillion. That is still astronomically expensive.

There is, however, a scenario that is always more expensive, no matter the cost of CDR. That is permanent global economic collapse caused by rapidly rising seas, flooded coast cities, desertification of the globe's key grain-exporting breadbaskets, colossal settlement-ravaging wildfires, proliferating disease, and attendant social breakdown. Unchecked, anthropogenic climate change threatens to become self-compounding, runaway, and unstoppable. James Hansen has even forecast the possibility of what he calls the "Venus syndrome," in which global warming over the course of two thousand years kills all life on earth.[15] The economic analog to Hansen's worst-case scenario, because it would "cost" everything, is therefore more expensive than the most expensive global campaign of CDR.

More optimistically, and correctly, J. W. Mason argues that thinking of climate adaptation in the zero-sum terms of cost and austerity misunderstands how the economy actually works. From a Keynesian perspective, investment creates economic growth, which in turn creates the resources for more investment. Or as Mason puts it, "The real resources for decarbonization will not have to be withdrawn from other uses. They can come from an expansion of society's productive capabilities, thanks to the demand created by clean-energy investment itself," and, we might add, from massive investment in CDR. Mason continues, "People rightly compare the scale of the transition to clean technologies to the mobilization for World War II. Often forgotten, though, is that in countries spared the direct destruction of the fighting, like the United States, war-time mobilization did not crowd out civilian production. Instead, it led to a remarkable acceleration of employment and productivity growth. Production of a liberty ship required 1,200 man hours [sic] in 1941, only 500 by 1944. These rapid productivity gains, spurred by the high-pressure economy of the war, meant there was no overall tradeoff between more guns and more butter."[16]

Extending the military analogy, Mason says that "the degree to which all war-time economies—even the United States—were centrally planned, reinforces a lesson that economic historians such as Alexander Gerschenkron and Alice Amsden have drawn from the experience of late industrializers: however effective decentralized markets may be at allocating resources at the margin, there is a limit to the speed and scale on which they can operate. The larger and faster the redirection of production, the more it requires conscious direction . . ."[17] Indeed, economic history reveals that massive economic transitions always require state coordination and subsidy, if not outright nationalization.[18]

In other words, the state could remove and store atmospheric CO_2 the same way that it currently builds dangerous and oppressive technologies like aircraft carriers and surveillance satellites. Society needs the utility of CDR deployed at a colossal scale as part of a multifaceted crash course of mitigation and adaptation. All indications are that only the decommodification of CDR and its deployment as a public utility can bring it to scale.

Society, the State, and Capital

CDR technology resists commodification and does not easily fit into capitalist social relations because its costs are too high and its benefits are too diffuse. Thus it must be treated as a global technology commons and deployed by governments as a public utility. This brings us to the second blockage: statephobia or state aphasia, as in the inability to see the state and comprehend its central role in modern life.

Let us begin by drawing a distinction between capital and capitalism. In classical Marxist terms, capital is a social relation. As Marx put it, "A cotton-spinning machine is a machine for spinning cotton. Only under certain conditions does it become capital. Torn away from these conditions, it is as little capital as gold is itself money, or sugar is the price of sugar."[19] In other words, "capital also is a social relation of production. It is a bourgeois relation of production, a relation of production of bourgeois society."[20] Only when the means of subsistence and the means of production are privately owned and used to command labor power in the goal of producing evermore exchange value are use values also capital.

Capitalism, on the other hand, is the whole ensemble of institutions that make up global society. Though dominated by capital, capitalism (or capitalist society) is not reducible only to the logic of capital. Throughout *Capital* vol. 1, Marx refers only to "the bourgeoisie" and "capital," never to "capitalism."[21] Society as a whole includes important countervailing forces, such as the state and public sector, social movements, precapitalist social formations and norms such as religions, and the whole noncommodified sphere of work referred to as "reproduction" or the love and care of families. In fact, capital (the social relationship of commodification and labor power exploitation) requires precapitalist and noncapitalist practices, institutions, and social formations to sustain it. Capital always depends on an "outside," as it were.[22]

Here, Karl Polanyi is essential. Polanyi showed that while all societies have embedded within them material processes that we would call "economic activity" and many include markets, no society, not even our own, has been totally governed by the rules of the market. As Polanyi put it in *The Great Transformation*, "The idea of a self-adjusting market implied a stark Utopia. Such an institution could not exist for any length of time without annihilating the human and natural substance of society; it would have physically destroyed man and transformed his surroundings into a wilderness."[23] To update that for the climate crisis, replace "wilderness" with "toxic wasteland of flooded coastal cities and desiccated agrarian interior."

For Polanyi, there

> was nothing natural about laissez-faire; free markets could never have come into being merely by allowing things to take their course. Just as cotton

manufactures—the leading free trade industry—were created by the help of protective tariffs, export bounties, and indirect wage subsidies, laissez-faire itself was enforced by the state. . . . The road to the free market was opened and kept open by an enormous increase in continuous, centrally organized and controlled interventionism. . . . The introduction of free markets, far from doing away with the need for control, regulation, and intervention, enormously increased their range.[24]

Polanyi argued that the transition to capitalism was so brutal and socially destabilizing that mitigating counteractions started immediately and organically—and came from reactionary forces as often as it did from progressive sectors like trade unions and socialist parties. Thus he wrote that while the "laissez-faire economy was the product of deliberate State action, subsequent restrictions on laissez-faire started in a spontaneous way. Laissez-faire was planned; planning was not."[25]

Polanyi correctly inverts the relationship between the state and capital. In Polanyi, we see that the state is always already deeply bound up with "the economy." It is in fact a crucial part of capital's life-support system. Thus while capital—the private sector on its own—cannot solve the climate problem, other elements within society acting through the state can, and must, take immediate steps to deploy the use value of CDR without waiting on social relations that are hostage to exchange values. In short, CDR must be treated like a modern commons or "public utility."

From the beginning of modern economics, even the most ardent advocates of laissez-faire have had to concede that certain goods and services cannot be managed by the market. Adam Smith hinted at this when groping toward the idea of "natural monopoly." Then in 1815, Malthus coined the term *natural and necessary monopolies*. Soon, many other writers were using *natural monopoly* to refer to "geographically fixed economic advantages in general."[26] In the middle of the nineteenth century, John Stuart Mill described natural monopolies as "those which are created by circumstances, and not by law."[27] For Mill, the risk associated with these natural monopolies was that they facilitated rent-seeking: "It is at once evident that rent is the effect of a monopoly."[28]

As Mill put it, "In the case of water-supply, there is virtually no competition. Even the possibility of it is limited to a very small number of individuals or companies, whose interest prompts them, except during occasional short periods, not to compete but to combine. In such a case, the system of private supply loses all that, in other cases, forms its recommendation."[29]

In the United States, Progressive Era economist Richard T. Ely first linked the idea of "natural monopoly" with the notion of "public utility." In his 1888 book *Problems of To-Day*, Ely listed key natural monopolies as "gas supply, street-car service, highways and streets, electric lighting, all railways, canals, bridges, lighthouses, ferries, docks, harbors, natural navigations, postal service, telegraphs and telephones."[30] Ely's contention was that "the regulation of these natural monopolies must be different from the regulation of commerce, agriculture, and manufactures" because "a natural monopoly . . . is excluded from the steady, constant pressure of competition."[31]

In other words, natural monopolies lead themselves to predatory rent-seeking. Ely argued that this situation left only two alternatives: "public control of private corporations, and public ownership with the public control which naturally springs from ownership." Ultimately, much crucial infrastructure was subject to considerable governmental control. Canals and railroads were built with vast public subsidies. Then during World War I, the federal government nationalized and reorganized the private railroad industry. Many cities built electrical grids or municipalized competing private grids.[32]

Even today, after forty-plus years of neoliberalism, there is no such thing as the free market. Unregulated capitalism is a myth; it has never actually existed. Nor could it. Capital is always bound up with the capitalist state. As Mariana Mazzucato has shown in her book *The Entrepreneurial State*, the modern capitalist state plays a central, guiding role in shaping technological innovation and deployment.[33]

In short, state economic action is not exotic and untested, as even many leftists and environmentalists believe; rather, it is ubiquitous yet veiled and denied. Step one in bringing CDR technology to scale is being realistic and honest about the dependence of capital upon the state.

Environment Making and Technology

Sorting out our relationship to CDR might require some clarification on the relationship between *Homo sapiens* and technology and a more historically informed understanding of the truly massive scale of our environment making. Lurking behind the specifics of CDR are larger questions about our role as a species. Should we attempt to restore a central feature of the global climate system? Is that not the height of hubris and bound for calamitous failure? Perhaps. But consider this: we have always used technology and have always been an environment-making species. Indeed, every species is. What we call "nature" or "the environment" is ultimately just the sum total of layer upon layer of organism-environment interactions. Every organism interacts with and impacts its environment. At the same time, every organism is always also part of the external environment of all other organisms. Environment making is what lifeforms do. Beavers need beaver ponds, but the creatures do not find their niche habitat; they make it by their prodigious and compulsive dam building.[34]

Or as Engels put it in the unfairly maligned and overlooked *Dialectics of Nature*: "Animals, as has already been pointed out, change the environment by their activities in the same way, even if not to the same extent, as man does, and these changes, as we have seen, in turn react upon and change those who made them. In nature nothing takes place in isolation. Everything affects and is affected by every other thing . . ."[35]

As a species, *Homo sapiens* are, Engels argued, the product of their own labor and technology of tools and fire: "The practical discovery of the conversion of mechanical motion into heat is so very ancient that it can betaken as dating from the beginning of human history . . . the making of fire by friction was the first instance of men pressing a non-living force of nature into their service."[36]

Further on, Engels noted how fire led to cooking, which led to profound physical transformations in the human body: "A meat diet contains in an almost ready state the most essential ingredients required by the organism for its metabolism. . . . The most essential effect, however, of a flesh diet was on the brain, which now received a far richer flow of the materials necessary for its nourishment and development, and which therefore could become more rapidly and perfectly developed from generation to generation. . . . With all respect to the vegetarians, it has to be recognised that man did not come into existence without a flesh diet" because cooked meat "further shortened the digestive process, as it provided the mouth with food already as it were semi-digested." And just "as man learned to consume everything edible, he learned [thanks to the warmth of fire] also to live in any climate. He spread over the whole of the habitable world, being the only animal that by its very nature had the power to do so."[37]

Modern environmental history has confirmed the importance and ubiquity of the human-fire relationship. As Stephen J. Pyne, the dean of fire studies, put it in one of the culminating books in his series on world fire culture, "there are no known peoples," except some Inuit and Yupik of the icebound far north, "who do not burn routinely" parts of their landscapes. All over the world foragers have

> recognized that berries, mushrooms, bracken, edible tubers like camas, and wild grasses flourished best on burned ground, that a light fire exposed acorns and chestnuts, that smoke deadened bees into a stupor that made honey accessible. Fishers recognized that torches attracted fish at night, when they could be easily speared or netted. Hunters saw that evening torches froze deer and geese, that flames could drive ungulates, that the fresh growth sprouting on old burns drew grazers, that fires flushed both elusive prey and dangerous predators from thickets . . . Regular burning, moreover, retained the desired habitat indefinitely. . . . Surely it is no accident that Artemis, the ancient goddess of the hunt—with an ancestry predating the Greeks—held a bow in one hand and a torch in the other.[38]

Very often this so-called broadcast burning creates quite fecund ecologies. Geoff Cunfer offers an impressive example in *Bison and People on the Northern Great Plains*: "On the northern plains, Village Indians developed a spring and fall burning cycle designed, in part, to manage bison . . . acre by acre, over several hundred years, Indians reworked the plains landscape" to enlarge the grassland, "converting millions of acres of forest to prairie." And by this environment making, "they increased bison populations to all-time highs, estimated at twenty-nine million by 1700."[39]

The larger point is this: we cannot retreat from our role as environment makers. Humans have always been remaking the planet. Unfortunately, under industrial capitalism, we do so as reckless, marauding somnambulants. We will destroy ourselves as a result, or we will become self-conscious, deliberate, life-producing, and life-enhancing environment makers.

Limits and Cynicism

A state-led crash program of CDR would only be meaningful within the context of a broader program of radical mitigation involving euthanizing the fossil fuel industry, a massive clean energy build-out, and robust adaptation efforts like coastal defense.

Pursued in isolation, CDR would do little to stabilize the climate system and could be used to justify inaction on the emissions reduction front. As it is, the promise of CDR is central to the myth of "clean coal." One could imagine the fossil energy sectors line if CDR were brought to scale: "No need to stop burning fossil fuel; we are stripping out so much CO_2!" Most current CCS is used by the oil industry.

Even if we could stabilize atmospheric CO_2 levels, there is still the problem of ocean acidification and spreading anoxia, both of which might bring down the human species. Despite potential problems with CDR, movements must demand that states embrace it and drive its mass adaptation. As Lackner put it, "Throwing a life-preserver to a drowning victim may not assure a successful rescue, but it is not a high-stakes gamble. Offering the life-preserver is preferable to withholding it, even though it might reduce the victim's incentive for learning how to swim."[40]

That sort of desperate optimism has much to recommend it. For one thing, it can help roll back the debilitating pessimism that comes with actually understanding climate science. Indeed, a major though largely unacknowledged problem among environmentalists is cynicism.[41] Anyone who comprehends the basic implications of climate science is forced to realize that climate change is unfolding in a nonlinear fashion in which the causes build while the effects lag. When the effects do begin to kick in, they will most likely do so with rapidly mounting intensity. Worse yet, half of all anthropogenic GHG emissions have occurred since 1990. Despite recent years of slowed emissions growth, the causes continue to build in exponential fashion. We are headed toward an extremely dangerous future. Knowing this, it is hard not to lapse into cynicism. Why struggle when the situation is so clearly hopeless? But if we do not struggle, then all is surely lost. In a book of conversations with international intellectuals, Fidel Castro addressed this problem: "If you knew that the world is going to last for ten years only, it is your duty to struggle and do something in those ten years. If somebody tells you, 'You can be certain that the planet is going to disappear and this thinking species is going to be extinct,' what would you do? Sit down and cry? I think we must struggle, and that's what we have always done. . . . that is what I suggest that we do, and not let ourselves be overtaken by pessimism."[42]

The failure of environmentalists to propose plans that credibly stand up to scrutiny in the face of climate science produces a lack of confidence, despair, and cynicism. Outright rejection of big technology feeds cynicism. For example, in an article rejecting CDR, an organizer with the (generally very worthy) Leap Manifesto wrote, "I myself have a very simple mantra: If Exxon is involved with something, it can't be good for our planet." ExxonMobil studies geology, but we don't reject geology just because they misuse it. Nor do we reject railroads simply because the Nazis used railroads. Similarly with CDR, the misuse of it by fossil fuel corporations is not a legitimate basis upon which to reject it. A radical approach to technology requires neither embrace nor rejection but rather thinking through the contradictions.

As I have said elsewhere, "Our mission as a species is not to retreat from, or to preserve, something called 'nature,' but rather to become fully conscious environment makers. Extreme technology under public ownership will be central to a socialist project of civilizational rescue, or civilization will not last."[43]

Notes

1   As suggested by Delina, this volume, chap. 8.
2   This technology is also referred to as carbon capture, usage, and storage (CCUS); it is also referred to as Carbon Capture and Sequestration (CCS).
3   For example, Anthony Karefa Rogers-Wright, "Carbon Capture and Storage: Sweeping Pollution under the Rug," *The Leap*, October 12, 2016, http://theleapblog.org/carbon-capture-and-storage-sweeping-pollution-under-the-rug/.
4   Peter Beaumont, "Scientists Suggest a Giant Sunshade in the Sky Could Solve Global Warming," *Guardian*, April 2018; "What Is Wrong with Solar Radiation Management?," ETC Group Briefing, March 24, 2017, http://www.etcgroup.org.
5   This study, though generally supportive of artificial fertilization of the ocean, found "the rainfall response may adversely affect water resources, potentially impacting human livelihoods" and "that changes in marine phytoplankton activity may lead to a mixture of positive and negative impacts on the climate." B. S. Grandey and C. Wang, "Enhanced Marine Sulphur Emissions Offset Global Warming and Impact Rainfall," *Scientific Reports* 5 (2015): 13055. For a critique of ocean fertilization, see Naomi Klein, "Geoengineering: Testing the Waters," *New York Times*, October 29, 2012.
6   Sam Holloway, "Carbon Dioxide Capture and Geological Storage," *Philosophical Transactions of the Royal Society A: Mathematical, Physical and Engineering Sciences* 365, no. 1853 (2007): 1095–1107.
7   Author's interview with Sigurður Reynir Gíslason, PhD, research professor at the Institute of Earth Sciences, University of Iceland, conducted on October 20, 2017, at CarbFix field injection site at Hellisheidi Geothermal Power Plant in Southwest Iceland.
8   Henry Fountain, "How Oman's Rocks Could Help Save the Planet," *New York Times*, April 26, 2018.
9   Klaus S. Lackner et al., "Carbon Dioxide Disposal in Carbonate Minerals," *Energy* 20, no. 11 (1995): 1153–1170.
10 David Nield, "Scientists Are Trapping CO_2 and Turning It into Stone Back Where It Belongs," *Science Alert* 23 (November 2016), http://www.sciencealert.com/scientists-are-again-turning-co2-into-solid-rock-to-fight-climate-change.
11 See the ClimeWorks website "Products" and "Customers" sections: http://www.climeworks.com.
12 Selin Ashaboglu, "Building Materials from Carbon Dioxide Emissions," *Architect*, August 29, 2017; Dom Galeon, "Revolutionary Carbon Capture Method Makes Building Materials Out of Emissions," *Futurism*, August 29, 2017; Christa Marshall, "Can CO_2 Be Captured and Sold?," *Scientific American*, December 21, 2012; Akshat Rathi, "Humanity's Fight against Climate Change Is Failing. One Technology Can Change That," *Quartz*, December 4, 2017.
13 Asbjørn Torvanger and James Meadowcroft, "The Political Economy of Technology Support: Making Decisions about Carbon Capture and Storage and Low Carbon Energy Technologies," *Global Environmental Change* 21 (2011): 305.
14 The symbol $ refers to U.S. dollars.
15 See chapter 10 in James Hansen, *Storms of My Grandchildren: The Truth about the Coming Climate Catastrophe and Our Last Chance to Save Humanity* (Bloomsbury: New York, 2009).

16   J. W. Mason, "When Output Is Limited by Demand, Action on Climate Change Doesn't
     Require Sacrifice," *International Economy*, Spring 2018, 20.
17   Mason, 20.
18   For an exploration of this point in the early American context, see Christian Parenti,
     Radical Hamilton: Economic Lessons from a Misunderstood Founder (New York: Verso,
     2020).
19   Karl Marx, "The Nature and Growth of Capital," in *Wage Labour and Capital* (Marxist
     Internet Archive, 1999 [1891]), 13, http://www.marxists.org/archive/marx/works/1847/
     wage-labour/.
20   Marx, "Nature and Growth."
21   Karl Marx, *Capital: A Critique of Political Economy, vol. 1*, 4th ed. (London: Penguin
     Books, 1976).
22   Jason W. Moore, *Capitalism in the Web of Life* (New York: Verso, 2015).
23   Karl Polanyi, *The Great Transformation: The Political and Economic Origins of Our Time*,
     rev. ed. (1944; repr., Boston: Beacon, 2001), 3.
24   Polanyi, 145, 146, 147.
25   Polanyi, 147.
26   Adam Plaiss, "From Natural Monopoly to Public Utility: Technological Determinism
     and the Political Economy of Infrastructure in Progressive-Era America," *Technology and
     Culture* 57, no. 4 (October 2016): 806–830.
27   John Stuart Mill, *Principles of Political Economy* (New York: D. Appleton, 1885), 255.
28   Mill, 270.
29   John Stuart Mill, *Public Agency or Trading Companies: Memorials on Sanitary Reform,
     and on the Economical and Administrative Principles of Water-Supply for the Metropolis, etc*
     (London: J. Gadsby, 1851), 20.
30   Richard T. Ely *Problems of To-Day: Discussion of Protective Tariffs, Taxation, and Monopo-
     lies* (Boston: Thomas Y. Crowell, 1885), 117.
31   Richard T. Ely, cited in Plaiss, "From Natural Monopoly to Public Utility," 812.
32   Gretchen Bakke, *The Grid: The Fraying Wires between Americans and Our Energy Future*
     (New York: Bloomsbury, 2017).
33   Mariana Mazzucato, *The Entrepreneurial State: Debunking Public vs. Private Sector Myths*
     (New York: Public Affairs, 2015).
34   New research indicates that some raptors in fire-prone landscapes, specifically brown
     falcons and black kites in Australia, also use fire to hunt. Though not yet recorded on film,
     eyewitness testimony by Australian firefighters, aboriginal people living in fire-prone parts
     of the outback, historical literature, and "an aboriginal ceremony where elders dress up as
     the Brown Falcons and Black Kites and move flaming sticks to spread fire in a symbolic,
     yet sacred, gesture" all suggest that the kites and falcons move fire as a hunting tactic. The
     birds are said to "pick up burning sticks and drop them in places they suspect delicious
     prey may be hiding. Then as lizards, snakes, frogs, and smaller birds scurry away from the
     flames, the predators are in perfect position to attack." See Kristen Schmitt, "Can Birds
     Actually Start Forest Fires?," *Audubon*, February 22, 2016.
35   Friedrich Engels, "The Part Played by Labour in the Transition from Ape to Man," in
     Works of Friedrich Engels (Moscow: Progress, 1934).
36   Friedrich Engels, "Chapter V Heat," in *The Dialectics of Nature* (Marxist Internet Archive,
     2001), http://www.marxists.org/archive/marx/works/1883/don/ch05.htm.
37   Friedrich Engels, "Part Played by Labour."
38   Stephen J. Pyne, *Vestal Fire: An Environmental History, Told through Fire, of Europe
     and Europe's Encounter with the World* (Seattle: University of Washington Press, 2012),
     30–31.
39   Geoff Cunfer and Bill Waiser, eds., *Bison and People on the North American Great Plains:
     A Deep Environmental History* (College Station: Texas A&M University Press, 2016), 13.
40   Klaus S. Lackner, "The Promise of Negative Emissions," *Science* 354, no. 6313 (2016): 714.

41 For an example of an eloquent embrace of pessimism, see Roy Scranton, *Learning to Die in the Anthropocene: Reflections on the End of a Civilization* (San Francisco: City Lights, 2015). While Scranton gets much correct, his political economy remains underdeveloped in that it fails to reveal the central, omnipresent role states have in creating and maintaining capitalism. Scranton does not endorse but does take at face value the late Thomas Shelling's ahistorical dismissals of state regulatory power.

42 Fidel Castro, *Fidel Castro Talks with Intellectuals: Our Duty Is to Struggle* (Havana: Instituto Cubano del Libro, 2016), 168–169.

43 Christian Parenti, "If We Fail," *Jacobin*, August 31, 2017, 130.

10

Planning the Planet

Geoengineering Our Way
Out of and Back into
a Planned Economy

ANDREAS MALM

When Paul Crutzen opened the floodgates for the geoengineering research enterprise in 2006, his rationale was simple: attempts to cut greenhouse gas (GHG) emissions "have been grossly unsuccessful."[1] So they have remained. Every year since 2006, the failure to restrain business as usual has become grosser. The case for geoengineering has gathered proportionate strength, all the wrong curves lending fresh credence to the notion that *we have to do something else*. Advocates of geoengineering—call them geoengineers for short—seem to be those most frustrated and panicky. "We're on a very unpleasant course, and we have to do something about it," says one of the experts featured in a promotion video for Harvard's Solar Geoengineering Research Program.[2] Geoengineers are not, of course, a homogenous group (even less a cabal of malicious pretenders to world rule), but the "we have to do something" argument is a recognizable common denominator. They cannot bear the thought of emissions continuing upward and are prepared to combat the warming by, literally, any means necessary. They have earned a reputation as the most impatient avant-garde who cannot stand idly by and demand to act on the anguish, sometime in the next few hours. But that reputation is undeserved. Alarm cannot be a sufficient condition for the advocacy of geoengineering. Everyone who knows something about the state of the climate is regularly racked by such feelings, but not everyone draws the conclusion that it's time to shoot the soot into the air or the CO_2 into the ground; in fact, there is little to suggest that the intensity of concern correlates with support for solar radiation management (SRM) or carbon dioxide removal (CDR) or some mix of the two. A second factor has to be present for someone to favor those kind of measures. The hidden premise, visible if you scratch the rhetoric a little, appears to be that geoengineers *rule out a planning of the economy*.

Any attempts to cut GHG emissions that would approach something like success would have to entail a significant degree of planning in the sense of governments assuming control over relevant parts of the economy—investment flows, consumption choices, trade, innovation, and so on—and steering them toward zero emissions at maximum speed. This is common knowledge, be it subconscious or conscious. Meaningful mitigation means massive, lengthy, single-minded interventions by the state into the day-to-day workings of the capitalist economy. A reminder of what happens in their absence came in July 2018, when the International Energy Agency (IEA) published its annual report on global trends in energy investment, a peephole into the actually existing economy as it operates outside of Intergovernmental Panel on Climate Change (IPCC) scenarios and computer simulations. In 2017, it turns out, more capital flowed into fossil fuels. Not only did investments in fossil fuels continue, but *their share relative to renewables expanded* so that 59 percent of all capital streaming into the production of energy ended up in oil and gas and coal. The agency expected the trend to last into 2018. Heeding the most basic findings of climate science would, of course, mean zero investment in *new facilities* for burning fossil fuels, but in the real world, the bulk of investment makes its way precisely to such additional ovens—and their owners are having a great time. The agency reported that oil and gas companies, responsible for the greatest investment boom, enjoy halcyon days, with the U.S. shale industry poised to generate across-the-board profits, which in turn will fuel new records in output; indeed, no less than 94 percent of global energy investment is financed "from capital incorporated into a company's balance sheet or from private individuals' own assets"—that is, from profits already made, a cycle otherwise known as capital accumulation. Oil and gas companies are so aflush in money that they can throw it into ever-greater capacity. But state-owned enterprises also account for a considerable amount of this—to put it mildly—misdirected investment, particularly in China; run like any other profit-maximizing corporations, they too accumulate their capital by extracting and combusting fossil fuels.[3] The general logic is summed up by an Australian team of scientists: "As long as it remains financially optimal, fossil fuel producers have incentives to exploit their reserves rapidly and continue exploration, *in direct conflict with other efforts to mitigate climate change.*"[4] As long as profits can be made, more fuel will be poured onto the planetary fire. Such is the grossness of the failure.

A break with this pattern would require that governments get a firm grip on the hundreds of trillions of dollars currently on their way into infrastructure worldwide and redirect them toward the things needed to stave off total climate breakdown.[5] An obvious start would be a ban on the construction of new fossil fuel facilities. That ban would then need to be progressively extended to the production and distribution of any of the three fossil fuel types: it is that simple and that comprehensive.[6] Whoever cares to listen to climate scientists will learn that "ambitious temperature targets"—1.5 or 2°C—can be reached solely through "a complete overhaul of the global economy."[7] One study of what it would take to stay below 1.5°C hints at the requisite aggressiveness of the interference: there would have to be "rapid forced closure of fossil-fuelled power plants" as well as "stringent enforcement of measures to reduce end-of-pipe emissions," like it or not.[8] Another high-profiled "roadmap for rapid decarbonization" does not mince words. There must be a "carbon law" mandating a halving of CO_2 emissions *every decade*, starting no later than 2020. It must apply "to all sectors and countries at all scales." Fossil fuels will have to be pulled out of the economy without hesitations and renewables pushed into every gap; in the 2030s, the last kerosene-fueled airplanes will disappear from the sky, the last internal combustion engines from the roads, all carbon-emitting steel and concrete from new buildings, and so on throughout the economy wherever energy is used. A "detailed road map" must be continuously updated every second year. By whom? The U.N. Framework Convention on Climate Change (UNFCCC), Johan Rockström and his team of scientists suggest, should be transformed into "a vanguard forum" that sets down the plan for the world as a whole.[9] It is Gosplan 2.0, or something very nearly like it.

The problem, of course, is that neither the UNFCCC nor its member states control the means of production that would have to be knocked out and rolled out, brought down and scaled up, rechanneled and repurposed for the one overarching aim of maintaining the planet in a habitable state: they are in private hands or in the hands of state-owned enterprises that behave just like any other capital. The latter segment could be redeployed with some ease. The former would first have to be *brought under public control*—not necessarily through direct ownership but unavoidably through instruments that would constrain the power of capitalists to dispose of their properties freely, be they taxes or tariffs, decrees or prohibitions, liquidations or procurements, discounts or subsidies—in whatever conceivable mix, an all-encompassing meddling with the economy. Precisely for this reason, discussions about climate mitigation often have something impotent and futile about them. *In the real world, capitalists do whatever they want* in relation to the science. For that to change, they would have to be reined in like a herd of bulls that must be prevented from running amok, but few dare to announce this loudly and fewer still have any concrete ideas for how to go about it.

We can, it follows, indulge in a bit of counterfactual history writing. Imagine that the world as a whole would have consisted of fully fledged command economies when climate science matured in the late 1980s and early 1990s. Imagine that these economies were plagued by sundry inefficiencies and bottlenecks, shortages and misallocations, but that they were governed by people prepared to accept the findings of the science. It is well known that command economies have one singular advantage: the state is well positioned to harness available resources in the service of one particular objective (such as winning a war, which is why even capitalist states have adopted planning in times of major wars).[10] We may conclude that attempts to cut emissions would have been rather more successful in this fictional world because, fundamentally, the means for affecting such cuts *would have already been in the hands of the governments* tasked with preventing dangerous anthropogenic interference with the climate system. They could have just turned some taps off and others on. Neither they nor the scientists would have needed to fumble for leverage in the economy. But in the real world, of course, the opposite happened. The birth of the UNFCCC coincided with the death of the very idea that the economy can or should be controlled for any purpose whatsoever.

Now, this line of reasoning will push all the buttons of the friends of free enterprise and markets. It flaunts the logic informing the climate deniers who believe that the science is a hoax perpetrated on freedom-loving people to pave the way for world tyranny. And there is a logic to their belief, however twisted it may be. It is the very same logic that predisposes some people to favor geo-engineering. They tend to articulate it in two ways: a planning of the economy such as is needed to meet ambitious temperature targets is either unlikely, difficult, or even *impossible*, or it is *undesirable*. Crutzen ended his article by stating that emissions cuts of the required magnitude would be the best outcome, but "currently, this looks like a pious wish."[11] Under *current* circumstances, that is, such a scenario seems unlikely to be realized. From that insight, one can either proceed to (1) investigate any potentials for making the scenario *more* likely by, for instance, trying to alter the balance of forces to the detriment of the vested interests or (2) examine entirely *different* methods for reducing the temperature on earth, and it is this second line of inquiry the geoengineers choose to pursue.

The Harvard promotion video opens with views of oil refineries, highways, coal trains, and container ports and the voice of Daniel Schrag saying that switching from fossil fuels to a different energy system would take "much more than a century" and cost "tens of trillions of dollars, and that's for the US alone."[12] This is not doable in the time required to minimize disaster. Likewise, to lay down *A Case for Climate Engineering*, David Keith begins by invoking the "inertia" of the economy. Very little, if anything, can be done about it. It is a fact before which we must bow. Scientists and activists not yet on board with geo-engineering "suffer from the persistent illusion that we can rapidly accomplish the deep structural change necessary to decarbonize our economy," that we can do something as herculean and presumptuous as "replacing a big chunk of the heavy infrastructure on which our society rests."[13] In *The Planet Remade*, Oliver Morton clears the stage for his preferred solution by advancing a number of arguments for why a rapid transition away from fossil fuels isn't practicable. The investments already made in such fuels are "huge." A shift to renewables won't come to pass by itself and would hence demand a lot of "pushing" from governments. Radical emissions cuts have no precedents in history (a criterion that clearly does not apply to the feasibility of geoengineering). Their probability is low and so is popular support for something like a carbon tax. In sum, "Fossil fuels are built into the foundations of the industrial and economic system, which means cutting emissions is hard—especially so since the costs of the cuts are concentrated on a powerful sector of the economy."[14] Capitalism as we know it rests on a bed of fossil fuels, and folding it would hurt those with power. So get the aerosol airplanes ready.

The premise for geoengineering—not so well hidden, after all—is the dismissal of a climate-planned economy as too challenging to contemplate, let alone advocate; it underpins the literature in the field. When four scientists in March 2018 published a proposal for geoengineering polar glaciers—building gargantuan seawalls and artificial islands around Greenland and Antarctica to stop the ice from sliding into the oceans—its raison d'être was the same. Swift and sweeping emissions cuts would save the ice, but they seem improbable. So let's focus on stemming the consequences instead (although these writers had the audacity to claim that "we address the problem [of sea level rise] at its source").[15] The average paper starts with observing that current emissions trajectories are way off track and immediately infers that "given this context," something like SRM must be considered.[16] The scramble for geoengineering after Paris, motivated by the glaring discrepancy between the sum of the national contributions and the officially promulgated targets, is but a specific version of this deduction. Given the context—that is, *taking the context as a given*—we have to try some other route. The entire enterprise is predicated on such surrender.[17]

This is, of course, partly a symptom of the crisis of political imagination: after 1989, the idea of breaking with the context of capitalism as it is was all but erased from the conceptual map. With the conclusive defeat of actually existing socialism, an "overwhelming heaviness" paralyzed the imagination, according to Enzo Traverso; the present order took on the guise of an immutable nature that is as hard to shake off as it would be to abolish the law of gravity—and "once capitalism is naturalized, to think of a different future becomes impossible."[18] Neoliberalism, according to Perry Anderson, "rules undivided across the globe: the most successful ideology in history."[19] This conjuncture is the indispensable setting for geoengineering. Indeed, it is hard to conceive of any other historical soil from which such a far-out prioritization could grow. Morton, for one, is aware of this: he knows of Fredric Jameson's dictum "it has become easier to imagine the end of the world than the end of capitalism," and *he elects to embrace that condition*.[20] Subjectively and objectively, geoengineering is an extreme expression of a—perhaps *the*—paradox of capitalist modernity. The structures set up by people are perceived as immune to tinkering, but there is hardly any limit to how natural systems can be manipulated. The natural becomes plastic and contingent; the social becomes set in stone.[21]

But geoengineers are not, it should now be clear, erstwhile revolutionaries who have at long last given up and resigned themselves to the intractable power of the enemy. People like Keith and Morton seem to think that the necessary planning is not only implausible and unattainable but also rather objectionable. Both men declare their devotion to capitalism and its helmsmen: "Must we fix capitalism in order to fix the climate?" asks Keith, cognizant of the alternative option. "Any serious argument in favour of this proposition," runs his answer, "must confront the fact that Western democracies have made enormous progress in managing environmental problems over the last half century." We *should* prefer technical to social fixes—that is, desist from transforming the economy and go for a technological solution—because capitalism is doing such an awesome job, including in the sphere of "environmental regulation."[22] Morton holds that geoengineering is a compelling prospect because it "can attract the interest and admiration of wealthier and more powerful men"—men of capital, among whom Bill Gates is the greatest; with his wife, he has saved the lives of as many people as Hitler, Stalin, and Mao have killed put together.[23] The virtues of this single billionaire are proportional to the worst compound of evil history has known. The descriptive, diagnostic argument—planning will not and cannot work—sits comfortably alongside a normative attachment to that which would have to be overhauled.

Business as usual is unchangeable *and*—or because it is—so very good: in the liberal super–best seller of 2018 *Enlightenment Now*, Steven Pinker offers his rendition of that combination of arguments (strikingly less informed, it should be pointed out, than Keith's or Morton's). He thinks that "the sacrifice needed to bring carbon emissions down by half and then to zero" would require "forgoing electricity, heating, cement, steel, paper, travel, and affordable food and clothing." The self-proclaimed guardian angel of enlightenment, reason, and science thinks that zero carbon emissions means shivering in the dark on a fixed spot without anything to read or wipe oneself with. He also thinks that "industrial capitalism launched the Great Escape from universal poverty in the 19th century and is rescuing the rest of humankind" in the twenty-first century. And he has warmed to geoengineering. From cloud nine ("everything is amazing"), he reports that people are more prone to accept the science of climate change "when they are reminded of the possibility that it might be mitigated by geoengineering than when they are told that it calls for stringent controls on emissions"—one more thing to celebrate. Having learned from Keith and Morton, Pinker now places his faith in bioenergy with carbon capture and storage (BECCS) as "climate change's savior technology."[24] The religious wording should be taken at face value.

All this seems to point to a foregone conclusion: geoengineering—in the air or on the land—is an escape from planning. The geoengineers strive to preempt any scenario of an economy commanded to respect the climate system. They seek to give capitalist anarchy a new lease on life. Nothing would then be easier than to write them off as peddlers of a red herring, strategists of a diversionary tactic designed to protect the ruling class and its freedom to trash the planet. That would be a rehearsal of the standard left position. Realities are, however, at a closer look, rather significantly more complex in the air as well as on the land.

Planning in the Air

Who would implement solar geoengineering? In what is arguably his most important paper since *A Case for Climate Engineering*, Keith assumes that it will be "a central planner." It will be a central planner who takes the "decisions about the implementation of SRM" in an all-inclusive effort "to maximize some measure of global welfare."[25] In this paper, Keith and his colleague Douglas G. MacMartin spell out the plan for a mode of solar geoengineering that is *temporary*, *moderate*, and *responsive*. Central planning is required to ensure all three qualities. Let us deal with the last two first. The injection of aerosols into the stratosphere has to stay within "moderate" bounds because the relief it offers from global warming does not increase linearly with the quantities sprayed; beyond some inflection point, the benefits of one more dose taper off, and the potential harms may well take over. Whatever negative consequences follow from the aerosols are likely to swell with the amount discharged.[26] The central planner must make sure that injections do not escalate perpetually, lest the atmosphere choke on the substance; being rational and benevolent, it will keep them within fairly tight limits.

A no less pressing duty is to stay "responsive." Keith and his colleagues cannot be accused of being blind wizards: they know that SRM would come with some very major risks and uncertainties. It is expected to reduce mean precipitation more than mean temperature on earth. That spells trouble, but on the other hand, rates of evaporation will also diminish when the planet cools off. A robust result of modeling is that SRM will have its largest cooling effect on the tropics, since that is where the blocked insolation would otherwise be most intense; the poles will be subject to a comparative "undercooling," meaning that temperature gradients between latitudes will even out. Now, the difference between hot tropics and cold poles drive much of the earth's weather systems, so SRM could severely disrupt their circulation. Holes in the ozone layer might grow. How the aerosols will interact with clouds is unclear.[27] The effects of more diffuse sunlight on photosynthetic productivity remain unknown; they could go either way. Centralized solar power plants might be starved of fuel, and agriculture might suffer from too little rain, too much rain, and/or too little sunlight: the economy down below will be susceptible to all kinds of distortions.[28] For these and any number of other reasons, the central planner must monitor how the real-time experiment unfolds and collect data at the highest possible resolution, process it, evaluate it, and feed it back into the planning. Benefits and harms must be weighed against each other and doses recalibrated accordingly. The planner might even have to call the whole thing off. Or it might switch to other substances, or slowly phase out the project, or extend it into "cocktail geoengineering" with a mix of techniques to fend off side effects, or decide on some other course of action. But whatever happens, "feedback is inevitable as the planner reacts to observations and discovers model errors."[29]

The imperative of planning is rather stronger here than in the realm of emissions cuts. *If solar geoengineering is to work, it has to be operated through central planning.*[30] Once the assumption of a central planner—one guided by reason and good intentions to boot—is articulated, all the worms come crawling out of the can. With a free market for SRM enterprises, one fleet could undo the work of another. It could spray at its own shelf in the atmosphere. It could ignore the latest information for any imaginable self-serving reason. A free-for-all could, according to Keith and MacMartin, induce "gross misuses of SRM or even war"—in short, uncontrollably spiraling calamities, something that does not figure in scenarios for centralized energy transitions.[31] Now, it is well established that sulfate aerosol injection is affordable and doable enough for "a wealthy individual" to launch on his own, or it could be done by a conglomerate of capitalists or a club of restless nations, but any such subglobal entity would then have the fate of the planet in its hands.[32] Few would feel comfortable with such an arrangement. There would rather have to be some fast reining in of the bulls. The literature now refers to this as the "free driver problem," which is opposite in structure to the free rider problem of emissions cuts (the temptation to benefit from the cuts of others while doing nothing). Essentially, pretty much any state and sufficiently wealthy corporation could start driving the airplane, with or without a green light from the rest of the international community.[33]

By its very nature, solar geoengineering nurtures a threat of anarchy, on which a lid must somehow be kept. One team pondering several different possibilities for the governance of SRM concludes that "a central hub for coordination" will probably be needed and spots a suitable candidate in the UNFCCC.[34] We're back at the vanguard forum again—only one whose activities would levitate above the ground. If planning for emissions cuts would aim at averting climate catastrophe by forcing agents to do some difficult things, planning for SRM would aim at averting climate catastrophe *and* geoengineering catastrophe by stopping agents from doing some easy things while ensuring that the things done are done the right way: with moderation and responsiveness.

But how would the central planner collect the data? And how would it be evaluated? Now, this makes for some further challenges. As Keith and MacMartin recognize, it is much easier to detect the result of an injection on average temperature than it is to pick out its impacts on one particular storm system, breadbasket, or monsoon season.[35] This is not because such lower-level impacts will be rare; on the contrary, by dint of the kind of radiative forcing it represents, solar geoengineering will affect the planet with far greater regional unevenness than will emissions cuts. It might fracture the dome of the global climate system into compartments that sail off in multiple directions, each with its own non-equilibrium, hysteresis, feedback loops, and consequent needs for adjustments.[36] According to MacMartin, Ricke, and Keith, "Choosing an appropriate level that balances different risks to the climate system will not be straightforward."[37] This problem has two components, the first of which is similar to the classical "local knowledge problem" of planned economies, as formulated by Friedrich Hayek: "The 'data' from which the economic calculus starts are never for the whole society 'given' to a single mind which could work out the implications, and can never be so given. The peculiar character of the problem of a rational economic order is determined precisely by the fact that the knowledge of the circumstances of which we must make use never exists in concentrated or integrated form, but solely as dispersed bits of incomplete and frequently contradictory knowledge"—a problem exacerbated by the infatuation with "statistical aggregates" that lumps everyone together and hides the nitty-gritty that matters.[38] This is roughly the problem any central planner of solar geoengineering would face, one difference being that it would concern the allocation of resources for survival. Hayek, of course, promoted a very straightforward solution. He suggested "decentralized planning by many separate persons," or leaving the decisions to the only one who can know about the local circumstances: "the man on the spot." But that solution is not available to solar geoengineering. We cannot have as many aerosol pilots as there are people affected by the weather. Besides being impracticable, it would be a recipe for cosmic chaos. Hayek urged the world "to dispense with the need of conscious control" and to trust the spontaneous conveyance of information through the price mechanisms of the market, but that model has zero application to SRM.[39] The central planner would have no other option than to take the bull by its horns and learn to manage the local knowledge problem as best it could.

The second component pertains not to knowledge but to politics: Whose preferences will determine what injections are dispensed? Whose interests will carry what weight? Ideally, there would be some mechanism in place to give the inhabitants of this planet equal say in the management of solar radiation. The team that wants to see the UNFCCC as the "central hub" also suggests a world-encompassing architecture of "focus groups," "deliberative workshops and exercises," national governments and U.N. agencies, and a "global forum" for civil society so as to guarantee the transmission of preferences and interests from the bottoms and peripheries to the apex—in other words, "a real participation in leadership of the interested masses themselves," to quote Leon Trotsky.[40] It would be an edifice of democratic planning, not in one country but on a world scale. Disparate wills could then be reconciled and the doses desired by humanity specified through collective deliberation, elections, assemblies, votes. Morton envisions a "Council of the Atmosphere" with members appointed through "global plebiscite."[41] The danger that some disaffected subaltern party would grab the pillars, bring down the whole SRM structure, and precipitate the dreaded termination shock would then be commensurately curtailed. If decision-making procedures allow for grievances to be aired from the start, perceived injustices will be addressed and the sunshade will go safe from sabotage.[42]

If this sounds like a rather ambitious plan for the planet, one could always lower the bar by letting one or a few powerful states take the initiative. One could wave on the first free driver. But if the adventurism and substitutionism of such a self-appointed vanguard aren't quickly reorganized into democratic planning, what are the odds that peace will prevail in the stratosphere? Those who lose out on the level of SRM established by the first driver might well consider it an act of willful aggression.[43] What if they hit back with countervailing measures and release potent, easily accessible GHGs—say, sulfur hexafluoride, some hydrofluorocarbon, a chlorofluorocarbon—to offset the effects of the aerosols? Or what if they start shooting down the aircraft? Keith and colleagues grapple more closely with these questions of threatening anarchy in a recent paper. They seek to prove that the odds for "counter-geoengineering" are high, on the grounds that it would be irrational. Counterattackers would have to calculate that benefits outweigh costs, and it is "unclear" that anyone would ever do so. Military intervention "would likely be viewed as a highly disproportionate response" by the targeted parties, and hence the chances for it are low. "An 'arms race' fought between escalating levels of" cooling with aerosols and warming with GHGs and back and forth and higher and higher "could pose large environmental risks to all people on the planet"—another reason this won't happen. No one in their right minds would uncork such a long-lasting and hyperpotent GHG as sulfur hexafluoride—it would be a "doomsday scenario."[44] But well, one might say, emitting a record thirty-three gigatons of carbon dioxide and investing another trillion U.S. dollars in infrastructure for combusting fossil fuels also look rather like a doomsday scenario and most certainly pose large environmental risks to all people on the planet, and yet this is exactly what happened in 2017, and on and on it goes. That is supposedly why we need geoengineering in the first place.

Here we encounter the self-contradictory, not to say schizoid, character of much of the geoengineering research enterprise. It is premised on the acceptance of a deeply irrational order as unchangeable and then leaps into heroic assumptions about the rationality with which the agents involved will handle the antidote to their own unreason. It's a bit like taking for granted that some-one with acute psychosis will treat himself in an optimally reasonable manner *plus* acquire just the right amount of food for a starving child now entrusted to him *and* judiciously deal with any unforeseen contingencies on their long journey together. One lesson of global warming so far would seem to be that the death drive of capitalist civilization should not be underestimated.[45] Yet Keith posits either a central planer pursuing "global welfare" with full rational-ity or bystanders reacting to the first driver with equally full rationality. In his contribution to this volume, Kevin Surprise points out that this first driver is likely to be—or very soon to be subjugated by—the U.S. military.[46] It would seem somewhat unwise to presume that this institution would rule the sky in the interests of humanity in general and the poor and vulnerable in particular. Nor can one postulate that the disadvantaged will sit idly by. Moreover, as Holly Jean Buck makes clear in her chapter, the eruption of post-truth and populist politics makes any presupposition about rational approaches to geoengineering seem rather slippery. A Donald Trump or Rodrigo Duterte with aerosols—the real-world leaders we have, beyond game theory—would not necessarily make safety and security drizzle down on the planet.

None of this, however, conclusively damns the enterprise. After all, a call for radical emissions cuts *also* departs from the assumption that things can be done in a more rational way than at present, pious as it may be. The high probability of a bad outcome is rarely a good reason to swallow it as a given—not, that is, if it derives from social relations. Politics doesn't work that way, or at least it shouldn't. It might still be the case that democratic planning of SRM *is* the way to go. What is needed, at the minimum, before any go-ahead is some recogni-tion of what kind of enterprise air-based geoengineering would be. Then what about land-based?

Planning on Land

Although it does not quite reach the dizzying heights of SRM sci-fi, the sec-ond main category of geoengineering schemes has its own fair share of risks and uncertainties. CO_2 injected into the ground might leak. BECCS projects might not so easily achieve a balance of negative emissions; initially, they might set off their own pulses of CO_2 if, for example, forests have to be cut down to make way for the biomass plantations. If the fuel has to be transported over long distances to power plants and the tubes of captured CO_2 to storage sites, emissions—the very thing that should be negated—might continue to plague the projects. The use of fertilizers on the plantations and energy-consuming processes for drying and grinding the biomass could have the same effect, potentially post-poning the point when the emissions turn negative by decades; we know from experiments with biofuels that high-flown expectations can be punctured by educated life cycle analyses.[47] Feedback would be needed here too. One cannot be confident that BECCS will *actually work* on a massive scale, and if it does, it might, somewhat like SRM, trigger other chain reactions in social and ecologi-cal systems—most obviously on the land.

Estimates of the land requirement for a worldwide deployment of BECCS that holds warming below 2°C by the year 2100 tend to hover in the vicinity of *all current cropland*.[48] Encore: another area as large as that presently devoted to cultivation would have to be found somewhere on earth for all those biomass plantations to have their intended effect. Where? Either in intact rainforests, grasslands, or savannahs—the final eradication of which would spell disaster for biodiversity and the carbon cycle and possibly other parts of the earth system—or on agricultural land already in use (this has been done before) where crops grown for people would have to be replaced with biofuels, with predictable results for cash-constrained households around the globe. Or one could *both* destroy rainforests *and* swap fuel for food. Either way, there's no exit from the trade-offs. The same goes for the freshwater resources that would have to be diverted—the plantations might be so thirsty for irrigation as to, again, *double* agricultural water consumption—and the gobbling up of fertilizers.[49] And all this voracity would occur in a late twenty-first-century world where land and water and phosphorous will be in *shorter supply* than at present, largely due, in the former two cases, to climate change itself. Add to that the imperatives of rewilding. Someone would have to make some very tough decisions.

That someone would need to attend closely to spatial zoning and location decisions: the distance between plantations, plants, and storage sites would have to be minimized and all the requisite transportation sourced from renewable energy. That someone would have to coordinate "a multi-national biomass supply chain," with plantations placed on the least damaging and most accessible sites in harmony with a no less extensive grid for renewable energy facilities—wind farms, solar arrays—with its very own claims on the land.[50] It would have to maintain "a detailed carbon accounting system" spanning countries and decades; provide "intelligent management" of the flow of resources, juggling consumption, production, distribution, and final storage; and find just the right "balanced array of BECCS activities" so as not to crash what remains of biodiversity.[51] Now, this doesn't sound very far from "the formulation of a unified economic plan, guidance of the work of the economic commissariats on the basis of the plan, verification of its fulfilment and, eventually, modification of the plan itself," in the words of the eighth congress of Soviets, convened in Moscow in 1920 to lay down the first-ever such plan; or "a single, centralized, global accounting," in the words of oppositional, left-communist pioneer planner Valerian Osinsky, who sought to establish an exchange of resources between town and countryside through a process of "mutual adjustment."[52] Needless to say, this isn't *exactly* what promoters of CDR and BECCS have in mind. But they do, wittingly or not, point toward some system of "book-keeping"—that is, with Karl Marx, "the supervision and the ideal recapitulation of the process" become evermore "necessary the more the process takes place on a social scale and loses its purely individual character."[53] Or can we perhaps leave it to the man on the spot?

Unlike SRM, CDR is in principle amenable to market mechanisms for the allocation of the resources in question. One option could be to construct a futures market for CDR. A buyer would go to that market to purchase, say, one ton of negative emissions. A seller would offer a contract to deliver those negative emissions at some specified future date. The seller would bank on a cheapening of CDR technologies in the meantime and expect to reap a profit from the difference between the costs at the future date and the price at the date of the transaction. This way, capital would flow to the CDR providers, who would be contractually obliged to deliver their services and strive to obtain maximum efficiency in the generation of negative emissions, in just the kind of spontaneously coordinated, decentralized web of price signals and local knowledge someone like Hayek would have loved. Today's financial capitalists would love it too.[54] But would it suck out any carbon?

Two researchers investigating the scheme conclude that a futures market for CDR would have an inbuilt tendency to favor those with the rosiest predictions for technological development: sellers would compete by presenting the lowest price per ton. Those counting on a gentle fall in CDR costs would automatically be outcompeted. There would be "Darwinian selection for delusion and denial among service providers. Firms making hubristic predictions may be the only ones able to attract buyers or investors. This poses extreme risks" because the anticipated technological advances may not have occurred when it's crunch time, or some other circumstance might have undermined the optimistic forecasts, and then the sellers will be incapable of delivering, go bankrupt, and quite possibly touch off a general collapse of the market, much like in 2008—with the difference that it would now be a scaffolding for reversing climate catastrophe coming down.[55] But it's not even certain that there would be much of a reversal involved. For buyers to flock to a CDR futures market, there would have to be some demand. For what? For negative emissions to *offset* the actual emissions produced by those buyers in the setup of this scheme—meaning that such market-driven CDR would only, in the best of cases, *compensate* for emissions that the purchasing capitalists cause elsewhere, not *draw down* carbon in any absolute sense.[56] The whole thing would have forfeited its purpose. And then we have not even touched on the ability of market-coordinated CDR to respect limits of land, freshwater, biodiversity, or any other kind of ecological constraint. The creativity of capitalist civilization when it comes to manufacturing markets out of thin air should not be underestimated, nor should its appetite for transgression.[57]

Who would set up the plantations? Who would build and equip the power plants? "No private entity would cover these costs without direct government intervention," two researchers assert.[58] In terms of resource mobilization, it would be another warlike undertaking for which capitalists seem to be their typical ill-prepared and indifferent selves; indeed, in its 2018 report, the IEA notices that only around 15 percent of the paltry $28 billion[59] earmarked for carbon capture and storage (CCS) projects since 2007 was ever spent "because commercial conditions and regulatory certainty have not attracted private investment."[60] An Osinsky with a mandate might be able to accomplish the mission. So might, perhaps, actively shepherded oil corporations, which are already endowed with most of what exists in the way of carbon storage technology; there is some potential for capital accumulation in this field.[61] Whether it would be rational to let such means of survival stay in private pockets is another matter. If market CDR is almost certain to mangle information about progress on the ground, it appears equally ill-equipped to deal with the *political* dimensions of a rush for BECCS: Who will give up their land for the plantations? Whose electricity will be fired by the biomass?

If the deep grooves of the world economy are anything to go by, land will be seized in the tropical South, from which the harvested biomass will be exported to the most advanced capitalist countries for burning. It would be ecologically unequal exchange on an epochal scale—and the proficiencies of capitalism in this regard are beyond doubt—bound to further fan the flames of resentment in the Global South: first you colonize us, then you wreck the climate, and then you colonize us again to make up for that wreckage and get some good fuel in the process.[62] A structure with a central hub in the UNFCCC and some real participation of the interested masses would be more fit to distribute the gains and losses of CDR in a just fashion—but then again, negotiations over historical responsibilities and burden sharing haven't exactly proceeded smoothly in the recent past. They would reach a new pitch on a negative planet. But there seems to be no way around that challenge imposed on the challenge, for even the most formidable plans for radical emissions cuts cited at the beginning of this chapter include a good deal of CDR *because too much has already been emitted.*[63] Faith in another savior technology further down the road should be treated just like that: there are no *rational* grounds for believing in a machine for carbon removal that transcends the restrictions of the material world any more than in a perpetual motion machine that overcomes the laws of thermodynamics.[64] Wherever we turn, we end up in the alley of planning.

Planning Three Economies in One

At a closer look, then, geoengineering offers nothing like an escape from a planned economy. The geoengineers do not table any remotely realistic proposals for extending the life and reign of capitalist anarchy. Instead, they suggest the construction of entire new economies—what we might call *geo*economies—that cry out for their own subordination to centralized planning. If the economy we now have serves to satisfy the material needs of people (a charitable interpretation), the two geoeconomies would have as their sole purpose to undo the worst damages from centuries of fossil fuel combustion. Each would have a distinct profile. The air-based geoeconomy would be thin and fast in that it would make modest demands on the resources of the economy below and have an all but instantaneous impact on the climate. The land-based would be thick and slow, as it would drill deep into the resource base of the economy and only gradually—over decades rather than months and years—leave its mark on temperatures. SRM is said to be orders of magnitude cheaper than normal mitigation, BECCS orders of magnitude more expensive.[65] If the former works flawlessly, it might not have to implicate itself in the foundations of the economy; if it flops, it would ruin them utterly; in the spectrum in between, the planning of it might attain some degree of intrusiveness. If the latter works, it is only because the planners have found a way to navigate the narrow straits of a resource-constrained world.

In fact, there is only one credible solution to the land dilemma—to wit, a complete termination of meat consumption. Two-thirds of all agricultural land today is occupied by pastures. *Here* is a reserve for biomass plantations, the only one whose requisitioning would not entail unpredictably ruinous knock-on effects on vital subsystems of the biosphere.[66] Freeing it up for the BECCS industry would require a cessation of the consumption first of meat and then of dairy products—mandatory global veganism, in plain language. (Lest the reader thinks this is based on some personal predilection, the present author admits to an addiction to halloumi, yogurt, eggs, and other lacto-ovo ingredients of the vegetarian Arabic and Iranian cuisines.) That would constitute a far-reaching intrusion into the metabolism of individual beings. But it would be hard to find some moral principle that permits them to go on consuming meat and dairy if it endangers the very habitability of the planet by denying CDR the only land on which it could sprout. It would be just as hard to justify exemptions for carnivores with purchasing power. States have banned narcotic substances, lethal weapons, and smoking in restaurants; if the prevention of breakdown and the stabilization of an unhinged climate are at stake, surely they could force their citizens to adopt what would incidentally be a healthier lifestyle. There has never been anything convenient about climate change, and it gets less convenient every hour. One might wish that no fossil fuels would ever have been burnt.

But the definitive proof that geoengineering opens no escape route from a planned economy is that at the end of the day, *it cannot substitute for radical emissions cuts*, and the geoengineers know this. The "temporary" in Keith's plan signifies that SRM will be used only for a while. If CO_2 continues to build up in the atmosphere, injections would have to escalate in tandem, with their own perilous side effects in tow: Oceans would be further acidified, irreversible climate change stacked into the sky for millennia to come. The central planner would have to plan so that SRM can be discontinued after one hundred years at the latest, a century during which emissions must be eliminated (and preferably turned negative).[67] There is a scientific consensus on SRM as, in the best case, buying some time and reprieve; it can never make mitigation redundant. While some might be *tempted* to put cuts on the shelf and bask in the filtered sunlight—the infamous "moral hazard"—it would fall to the lot of the planner to uphold the discipline and conduct the spraying as a mere supplement.[68] BECCS is likewise a pointless exercise in the absence of radical emissions cuts. It might be used to neutralize residual, unavoidable emissions from, say, rice cultivation or irreplaceable kerosene-fueled flights (if there are any such), but it cannot achieve negative emissions if positive emissions do not approach zero, not to mention continue unabated. Success in the geoeconomies pivots on what happens in the substrative economy. Given the very real and abiding grossness of the latter, the ultimate scenario for a concerted rescue operation might be *a combination of all three*—mitigation, management, and removal—so as to bring down temperatures and CO_2 concentrations to enduringly safe levels. That would leave us with *three economies to plan in one*.

Some geoengineers would perhaps adduce the unlimited potentials for collection of information thanks to big data, algorithms, artificial intelligence, and other technological breakthroughs. With those in hand, we can plan the two geoeconomies swimmingly, without serious deficits or mismatches of knowledge. Marxist advocates of planned economies have made the same argument for some time now.[69] Twentieth-century socialist economies did not have the advanced computers, but we can use them to overcome Hayek's local knowledge problem (as well as democratic deficits and other well-known drawbacks). This kind of argument pertains to the means available for successful planning, but it does not in any way reduce the need for it. The marvels of digital technology might help us make the challenge of triple planning a little easier; they will not remove it as such.

If the tech-case for effective geoengineering would echo positions from the farthest left, however, most of the left and the climate movement would balk at a scenario of combining mitigation with management and removal. SRM and even CDR still evoke haphephobia. Some cling to the belief that geoengineering will be "infinitely more dangerous than climate change" on the basis of some poorly examined axioms; worse, this position may now imply complacency with the very real and terrifying dangers of climate change, already activated with 1°C of warming and stretching toward a truly infinite horizon of cataclysm.[70] (These words are written in an apartment in Sweden as the country suffers its worst season of wildfires in modern history, with forests from the Arctic North to the southern coast reduced to ashes and dozens of settlements evacuated, on top of a drought that has devastated the agricultural sector—all this in a land that was supposed to be among the last to feel the heat. And this is nothing compared to what goes on in Iran or Iraq or Pakistan.)

The left too has its taboos to give up. They include deep-seated hostility to nuclear power, aversion to centralized solar power plants and other large-scale infrastructure, and blanket rejection of any talk of geoengineering. The latter is no longer a tenable position, if it ever was; not long after climate change emerged as a problem, it became clear that CO_2 concentrations would need to be reduced to, for example, 350 ppm, as proposed by 350.org. CDR must be part of any toolbox for the remediation of global warming (although one can imagine it in other forms than BECCS, such as soil carbon sequestration, afforestation, and rewilding).[71] As for SRM, it can hopefully be avoided—kept outside the box—but a position that a priori excludes any and every scenario where its deployment would be preferable to unalleviated climate change is guilty of a whole series of underestimations. Taking climate change seriously means reexamining ingrained notions and striking out as boldly as the situation demands, even if it includes listening to someone like Keith. He and his peers may have the most odious ulterior motives for their enterprise, such as an adoration for capitalism, their most intimate partner, who constantly nourishes and enriches their lives; or Promethean hubris; or boyish excitement about the power of contraptions—"it is hard not to delight in these newfound tools"; "some see beauty in these new forms of engineering."[72] Morton in particular invests all the utopian energies vacated from the social sphere in the brave new world of aerosols. As psychopolitical bourgeois pathologies go, this must count as one of the more remarkable.

Geoengineers, then, might be animated by deep drives in which no one on the left or in the climate movement would wish to share. But Marxists can discern in their schemes the tendency for objective socialization of the means of—in this case—survival, poised to burst their capitalist integument asunder (or else burst the planet asunder). Furthermore, revolutionary Marxists have always prepared *to inherit the catastrophes of capitalism.* That is what the Bolsheviks had to do in war-torn Russia, groping their way toward planning as the country went up in flames around them with no manual to follow, no template for how to mix market mechanisms and state control in the right proportions, and no guarantee that anything they did would work. It ended badly, of course. The post–1989 crisis of the political imagination has something to do with how the planned economies of the Soviet Union and its satellite states panned out (which implies, if one is so inclined, that the geoengineering research enterprise can in the last instance be deduced from the degeneration of the Russian Revolution). Speaking of gross failures, these cannot, obviously, be repeated, but as it happens, the Stalinist experience was shadowed by a short century of disputes and conversations about how to build democratic planning, from the feverish debates in the wake of October—in which Lenin himself warned that "a complete, integrated real plan is a bureaucratic utopia"—to the discussions about "market socialism" after the fall of the wall.[73] Perhaps there is reason to revisit that tradition. In a rapidly warming world, more or less everyone—the denialists, the geoengineers, the champions of cuts—seems to be convinced that climate change can only be tackled by planning. We might just as well be honest about it.

Notes

1   Paul J. Crutzen, "Albedo Enhancement by Stratospheric Sulfur Injections: A Contribution to Resolve a Policy Dilemma?," *Climatic Change* 77 (2006): 211–212.

2   Richard Zeckhauser in "A New Tool to Address Climate Change," video, Harvard University, accessed December 21, 2018, geoengineering.environment.harvard.edu.

3   International Energy Agency (IEA), *World Energy Investment 2018* (Paris: IEA, 119); see also, for example, 12–14, 119–120, 125, 132, 138.

4   Saphira A. C. Rekker et al., "Comparing Extraction Rates of Fossil Fuel Producers against Global Climate Goals," *Nature Climate Change* 8 (2018): 489 (emphasis added).

5   Ilmi Granoff, J. Ryan Hoghart, and Alan Miller, "Nested Barriers to Low-Carbon Infrastructure Investment," *Nature Climate Change* 6 (2016): 1065–1071.

6   Fergus Green, "The Logic of Fossil Fuel Bans," *Nature Climate Change* 8 (2018): 449–450.

7   Oliver Geden, "Politically Informed Advice for Climate Action," *Nature Geoscience* 11 (2018): 382. Cf., for example, Joeri Rogelj et al., "Scenarios towards Limiting Global Mean Temperature Increase below 1.5°C," *Nature Climate Change* 8 (2018): 327.

8   Detlef P. van Vuuren et al., "Alternative Pathways to the 1.5°C Target Reduce the Need for Negative Emission Technologies," *Nature Climate Change* 8 (2018): 396, 392.

9   Johan Rockström et al., "A Roadmap for Rapid Decarbonization," *Science* 355 (2017): 1269–1270.

10 Stalinist Soviet Union beat everyone else in this game. In World War II, no other combatant country reached its levels of total mobilization of the resources of the economy—75 percent of net national product already in 1942, up from 20 percent in 1940. And these were, of course, the two years when the Soviet economy was ravaged by Nazi invasion. The United Kingdom went from 48 percent to 54 percent in the same period. Mark Harrison, "Resource Mobilization for World War II: The U.S.A., U.K., U.S.S.R., and Germany, 1938–1945," *Economic History Review* 41 (1988): 184. "In the years from mid-1941 to mid-1944 Soviet resources were employed in the cause of Germany's

military defeat with far greater intensity than those of the United Kingdom or North America." Harrison, 191. Harrison stresses, however, that this achievement was not solely the result of top-down planning: "Even in the first, comparatively leaderless, days" after the onset of Operation Barbarossa "the conversion and mobilization of the economy for war production was in full swing. People knew what they were supposed to do and did it without having to be told directly.... In summary, there were two elements in Soviet economic resilience in 1941–2. One was the capacity of Soviet leadership for high-level initiative and individual improvisation, enforced by decrees and dictatorial powers, in the face of emergency. The other was the popular response from below." Harrison, 178.

11   Crutzen, "Albedo Enhancement," 217.

12   Zeckhauser, "New Tool to Address Climate Change."

13   David W. Keith, *A Case for Climate Engineering* (Cambridge, Mass.: MIT Press, 2013), xii, 29, 31. Strangely—or not so strangely—this pessimism is combined with great optimism about the benefits of growing wealth: "Our wealth now grows much faster than our carbon emissions, so it gets steadily easier to cut our carbon footprint." Keith, 164.

14   Oliver Morton, *The Planet Remade: How Geoengineering Could Change the World* (London: Granta, 2015), 8, 13, 145; see further 4, 11, 19, 265, 346–347.

15   John C. Moore et al., "Geoengineer Polar Glaciers to Slow Sea-Level Rise," *Nature* 555 (2018): 303–305 (quotation from 304).

16   Douglas G. MacMartin, Katharine L. Ricke, and David W. Keith, "Solar Geoengineering as Part of an Overall Strategy for Meeting the 1.5°C Paris Target," *Philosophical Transactions of the Royal Society A* 376 (2018): 2.

17   Cf. the argument made in Ryan Gunderson, Brian Petersen, and Diana Stuart, "A Critical Examination of Geoengineering: Economic and Technological Rationality in Social Context," *Sustainability* 10 (2018): 11, 14.

18   Enzo Traverso, *Left-Wing Melancholia: Marxism, History, and Memory* (New York: Columbia University Press, 2016), 4, 57.

19   Perry Anderson, "Renewals," *New Left Review* 2, no. 1 (2000): 13.

20   Morton, *Planet Remade*, 310; cf., for example, 220.

21   Cf., for example, Andrew Biro, "Introduction: The Paradoxes of Contemporary Environmental Crises and the Redemption of the Hopes of the Past," in *Critical Ecologies: The Frankfurt School and Contemporary Environmental Crises*, ed. A. Biro (Toronto: University of Toronto Press, 2011), 4–5.

22   Keith, *Case for Climate Engineering*, 143–147.

23   Morton, *Planet Remade*, 27, 353.

24   Steven Pinker, *Enlightenment Now: The Case for Reason, Science, Humanism, and Progress* (New York: Viking, 2018), 141, 364, 283, 382, 151.

25   David W. Keith and Douglas G. MacMartin, "A Temporary, Moderate and Responsive Scenario for Solar Geoengineering," *Nature Climate Change* 5 (2015): 201.

26   Keith and MacMartin, 201–202; cf. MacMartin, Ricke, and Keith, "Solar Geoengineering," 2.

27   MacMartin, Ricke, and Keith, "Solar Geoengineering," 4–6.

28   For an overview, see Peter J. Irvine et al., "An Overview of the Earth System Science of Solar Geoengineering," *WIREs Climate Change* 7 (2016): 815–833.

29   Keith and McMartin, "Temporary, Moderate and Responsive Scenario," 204. "Cocktail geoengineering": Long Cao et al., "Simultaneous Stabilization of Global Temperature and Precipitation through Cocktail Geoengineering," *Geophysical Research Letters* 44 (2019): 7429–7437.

30   Or, as Keith observed already, "geoengineering seems to demand centralized control." Keith, *Case for Climate Engineering*, 153.

31   Keith and McMartin, "Temporary, Moderate and Responsive Scenario," 201.

32   Janos Pasztor, Cynthia Scharf, and Kai-Uwe Schmidt, "How to Govern Geoengineering?," *Science* 357 (2017): 231.

33 See, for example, Anthony Harding and Juan B. Moreno-Cruz, "Solar Geoengineering Economics: From Incredible to Inevitable and Half-Way Back," *Earth's Future* 4 (2016): 569, 573.

34 Simon Nicholson, Sikina Jinnah, and Alexander Gillespie, "Solar Radiation Management: A Proposal for Immediate Polycentric Governance," *Climate Policy 19, no. 3*: 327; cf. 329.

35 Keith and McMartin, "Temporary, Moderate and Responsive Scenario," 205.

36 This point is argued forcefully by Mike Hulme, *Can Science Fix Climate Change? A Case against Climate Engineering* (Cambridge, U.K.: Polity, 2014), 48–53.

37 MacMartin, Ricke, and Keith, "Solar Geoengineering," 8.

38 Friedrich A. Hayek, "The Use of Knowledge in Society," *American Economic Review* 35 (1945): 519, 524, 521, 527.

39 Hayek, 521, 527.

40 Nicholson et al., "Solar Radiation Management," 324, 326, 330; Leon Trotsky, *The Revolution Betrayed*, rev. ed. (1937; repr., Mineola, N.Y.: Dover, 2004), 51.

41 Morton, *Planet Remade*, 373.

42 Andy Parker and Peter J. Irvine, "The Risk of Termination Shock from Solar Geoengineering," *Earth's Future* 6 (2018): 462.

43 Olaf Corry, "The International Politics of Geoengineering: The Feasibility of Plan B for Tackling Climate Change," *Security Dialogue* 48 (2017): 306. "Any unusual weather extremes may be blamed on the SRM deployment." Keith and MacMartin, "Temporary, Moderate and Responsive Scenario," 205.

44 Andy Parker, Joshua B. Horton, and David W. Keith, "Stopping Solar Geoengineering through Technical Means: A Preliminary Assessment of Counter-Geoengineering," *Reviews of Geophysics* (2018): 1058–1065. The most disingenuous argument is perhaps the following: "Contrary to the common assumption that the ability to engage in solar geoengineering would be widely distributed among states, practical requirements related to delivery infrastructure, technical capacity, and ability to withstand external pressure would likely mean that SRM capabilities would be limited to major powers or coalitions." Parker, Horton, and Keith, 1062. Such an inherent monopoly is a laudable property of SRM, in this context. But in *A Case for Climate Engineering*, Keith himself makes exactly the opposite argument: that SRM is so fabulously cheap and easy that pretty much anyone can do it. The technologies "could be produced by a surprising number of countries within a decade.... The technology necessary to build an aircraft that could deliver payloads to the lower stratosphere is much less sophisticated than that necessary to build an advanced fighter aircraft. The job can be effectively accomplished with quite conventional jet aircraft technology," a technology that virtually all countries in the world currently possess. Keith, *Case for Climate Engineering*, 153–154. In the article "Stopping Solar Geoengineering," Keith et al. seek to demonstrate that countergeoengineering is implausible, and hence they claim that SRM is out of reach for all but a few; in the book *A Case for Climate Engineering*, Keith aims to show that geoengineering is feasible and asserts that SRM is open for everyone. The tendentiousness is conspicuous.

45 "It's only 4 degrees, it's only 4 degrees / I want to see this world, I want to see it boil / I want to see this world, I want to see it boil / I want to see the fish go belly-up in the sea / And all those lemurs and all those tiny creatures / I want to see them burn, it's only 4 degrees / And all those rhinos and all those big mammals / I want to see them lying, crying in the fields / I want to see them, I want to see them burn." "4 Degrees," track 2 on Anohni, *Hopelessness*, Secretly Canadian, 2016. More accurate modelling would be hard to find.

46 Surprise, this volume, chap. 13.

47 For an overview of these and other problems, see Mathilde Fajardy and Niall Mac Dowell, "Can BECCS Deliver Sustainable and Resource Efficient Negative Emissions?," *Energy & Environmental Science* 10 (2017): 1389–1426.

48 See, for example, Michael Obersteiner et al., "How to Spend a Dwindling Greenhouse Gas Budget," *Nature Climate Change* 8 (2018): 8; Christopher B. Field and Katharine J.

Mach, "Rightsizing Carbon Dioxide Removal," *Science* 356 (2017): 707; and Kate Dooley, Peter Christoff, and Kimberley A. Nicholas, "Co-producing Climate Policy and Negative Emissions: Trade-Offs for Sustainable Land-Use," *Global Sustainability* 1 (2018): 1.

49 Vera Heck et al., "Biomass-Based Negative Emissions Difficult to Reconcile with Planetary Boundaries," *Nature Climate Change* 8 (2018): 151–155; Holly Jean Buck, "Rapid Scale-Up of Negative Emissions Technologies: Social Barriers and Social Implications," *Climatic Change* 139 (2016): 158. See further, for example, Vera Heck, Jonathan F. Donges, and Wolfgang Lucht, "Collateral Transgression of Planetary Boundaries Due to Climate Engineering by Terrestrial Carbon Dioxide Removal," *Earth System Dynamics Discussions* 7 (2016): 783–796. On the lessons of the early twenty-first-century biofuels boom for BECCS, cf. Matthias Honegger and David Reiner, "The Political Economy of Negative Emissions Technologies: Consequences for International Policy Design," *Climate Policy* 18 (2018): 311–312.

50 Fajardy and Mac Dowell, "Can BECCS," 1391; cf. 1397. For an argument emphasizing the land claims of renewables, see Matthew T. Huber and James McCarthy, "Beyond the Subterranean Energy Regime? Fuel, Land Use and the Production of Space," *Transactions of the Institute of British Geographers* 42 (2017): 655–668.

51 Glen P. Peters and Oliver Geden, "Catalysing a Political Shift from Low to Negative Carbon," *Nature Climate Change* 7 (2017): 621; Fajardy and Mac Dowell, "Can BECCS," 1411; Paul C. Stoy et al., "Opportunities and Trade-Offs among BECCS and the Food, Water, Energy, Biodiversity, and Social Systems Nexus at Regional Scales," *BioScience* 68 (2018): 109.

52 Quoted in Silvana Malle, *The Economic Organization of War Communism, 1918–1921* (Cambridge: Cambridge University Press, 1985), 312, 297.

53 Karl Marx, *Capital*, vol. 2 (London: Penguin, 1992), 212.

54 D'Maris Coffman and Andrew Lockley, "Carbon Dioxide Removal and the Futures Market," *Environmental Research Letters* 12 (2017): 4.

55 Coffman and Lockley, 5; cf. 7; and cf. Buck, "Negative Emissions Technologies," 161.

56 Coffman and Lockley, "Carbon Dioxide Removal," 4. Cf. Honegger and Reiner, "Negative Emissions Technologies," 311.

57 Another kind of potential CDR market is investigated in Andrew Lockley and D'Maris Coffman, "Carbon Dioxide Removal and Tradeable Put Options at Scale," *Environmental Research Letters* 13 (2018): 1–11.

58 Honegger and Reiner, "Negative Emissions Technologies," 313. Cf. 308.

59 The symbol $ refers to U.S. dollars.

60 IEA, *World Energy Investment 2018*, 17.

61 For more on this scenario, see Holly Jean Buck, "A Best-Case Scenario for Putting Carbon Back Underground," *Science for the People*, July 24, 2018, http://magazine
.scienceforthepeople.org/best-case-scenario-carbon-underground/.

62 See Peters and Geden, "Catalysing a Political Shift."

63 See Van Vuuren et al., "Alternative Pathways"; and Rockström et al., "Roadmap for Rapid Decarbonization."

64 For one promising technology that might well outperform BECCS and will certainly have its own limitations, see Jinyue Yan, "Negative-Emissions Hydrogen Energy," *Nature Climate Change* 8 (2018): 560–561; and Greg H. Rau, Heather D. Willauer, and Zhiyong Jason Ren, "The Global Potential for Converting Renewable Electricity to Negative-CO_2-Emissions Hydrogen," *Nature Climate Change* 8 (2018): 621–625.

65 For the cheapness of SRM, see Keith, *Case for Climate Engineering*, 7, 100; and Keith and MacMartin, "Temporary, Moderate and Responsive Scenario," 205. For the expensiveness of BECCS, see Honegger and Reiner, "The Political," 309.

66 See, for example, Lena R. Boysen, Wolfgang Lucht, and Dieter Gerten, "Trade-Offs for Food Production, Nature Conservation and Climate Limit the Terrestrial Carbon Dioxide Removal Potential," *Global Change Biology* 23 (2017): 4303–4317.

67 Keith and MacMartin, "Temporary, Moderate and Responsive Scenario," 204–205; cf. MacMartin, Ricke, and Keith, "Solar Geoengineering," 2, 12.

68 Cf., for example, Nicholson et al., "Solar Radiation Management," 324–325. There are, however, occasional slips of the tongue: Keith and colleagues can suddenly speak of the importance of measuring the cost-effectiveness of SRM "because the comparison is to other means of achieving the same result," which implies substitution. David W. Keith, Gernot Wagner, and Claire L. Zabel, "Solar Geoengineering Reduces Atmospheric Carbon Burden," *Nature Climate Change* 7 (2017): 618. Morton, who is not a scientist, succumbs to the temptation: "If research on geoengineering pays off in providing a way of slowing the warming that can be deployed in a satisfactorily safe and just manner, then reduced interest in mitigation could be rational and morally justified." Morton, *Planet Remade*, 160. Such is the dream of escape.

69 Notably Paul W. Cockshott and Allin Cottrell, *Towards a New Socialism* (Nottingham, England: Russell, 1993).

70 James Rodger Fleming, quoted in Gunderson et al., "Critical Examination of Geoengineering," 7.

71 For afforestation and rewilding, see, for example, Troy Vettese, "To Freeze the Thames," *New Left Review* 2, no. 111 (2018): 63–86.

72 Keith, *Case for Climate Engineering*, 173–174; Morton, *Planet Remade*, 372 (cf. 139, 337–338, 345).

73 Lenin in Malle, *Economic Organization of War Communism*, 313.

11

Provisioning Climate

An Infrastructural Approach
to Geoengineering

ANNE PASEK

The climate crisis is a techno-political one. This point has been emphasized in recent Marxist and anticolonial scholarship, which examines the historical carbonization of industrial production through particular energy technologies not only as a question of economic development but also as a decisive means by which political subjugation and contestation were waged.[1] The future and massive decarbonization of production, a necessary precondition of any plausible climate mitigation effort, poses these relations anew and with them the great potential for redistributions of power in both its political and technical senses.[2] As many of the contributions in this book forward, it is both possible and necessary to map geoengineering into the critical frameworks of this very discussion, assessing the distinct strategic affordances of solar radiation management (SRM) and carbon dioxide removal (CDR) within present and future struggles over the control of industrial and geophysical systems. It matters who controls geoengineering systems, to what ends, and on whose dime.

However, as this chapter argues, it is also possible and necessary to consider how such dynamics in turn challenge what we mean by technology and politics and whether these terms are adequate to the task of assessing the benefits and risks of large climatic interventions in a warming world—one where the reproduction of global ecological systems, not only the production of goods or national wealth, has come to the fore as a decisive question in the distribution of power. As we engage in the work of analyzing and framing speculative technologies, I hold that techno-politics too must be subject to speculation and debate. We must question how to best frame geoengineering within different analytic vocabularies and attend to the ways in which geoengineering may generate new social formations in turn.

The distinct scalar and territorial character of geoengineering can be difficult to reconcile within traditional models of geopolitics. I begin this chapter by assessing the small but growing body of literature within the humanities and social sciences that considers how such a reconciliation may be attempted, for better or for worse. Noting the evident contradictions within the scholarship so

163

far, I intervene by proposing that this impasse seems to follow from a restricted model of technology as a category of analysis.

Instead of framing geoengineering as a fixed tool for the making of climate, I argue that it is more accurate to the proposed technical systems and more productive for our political discourse to approach it as an infrastructural formation. The gains of this shift in our thinking are threefold. Infrastructural analysis allows for pastoral divisions between nature and technology to be replaced with more nuanced and productive continuities; it allows for new climatic social formations to be more clearly assessed, in both their positive and their negative potentials; and it further reorients reactionary perspectives on technology, labor, and repair to be qualified and rearticulated within a social reproduction framework.

This argument is informed by a critical science and technology studies perspective as well as feminist technoscience provocations and theory. These literatures contain a diverse, and often contradictory, set of political and disciplinary traditions. The discussion that follows, resultantly, is broad and evocative rather than conclusive. These observations are offered toward the development of improved future debates on the character and demands of progressive geoengineering political strategies. I suggest where we might look for new tensions, claims, and coalitions rather than programmatically arguing for when and how to act.

Recent Attempts to Reconcile Geoengineering and Politics

To geoengineer is to initiate a fundamentally different kind of relationship to territory and ecology than previous attempts at environmental management or geopolitical control. It implies and enacts a shift from the governance of planetary surfaces to the governance of planetary systems.[3] Such scales of action, combined with potentially nonlinear and unpredictably distributed secondary effects, are likely incompatible with Westphalian notions of national sovereignty or forms of precautionary risk management that do not engage in counterfactual deliberation. To commit to geoengineering requires an orientation toward planning and intervention that makes the philosophical implications of the Anthropocene both banal and newly urgent—changes to the earth system become an epochal configuration that must be intentionally managed and mitigated rather than an abstract condition into which we are thrown by history.[4] It is thus both the scale of geoengineering's programmatic breadth and its potential consequences that attract the attention of critics and proponents alike.

Assessments of what this might mean for future sociopolitical formations, however, have thus far been both strident and contradictory. Early commentators of geoengineering note that these material affordances call for the theorization of a new kind of geopolitics, one that, as Simon Dalby and Nigel Clark summarize, might more directly address geology than the nation-state.[5] Similarly, Kathryn Yusoff suggests that the material and political affordances of geoengineering's agential scope and scale must be matched by a similar intellectual move beyond "flat earth" international relations.[6] In both the transnational maneuvers implied by such geophysical interventions and the risk of their potential market enclosures, geoengineering suggests a particular urgency to global political calculations and the potential need for new forms of planetary and popular deliberation.

However, beyond these invocations of sea changes, figurative and literal, little work has been done that attempts to fully apply a geological geopolitics to an analysis of the technological forms that are said to bring this political concept into being. Instead, the work that explicitly addresses the politics of geoengineering systems as such tends to revert back to traditional categories of governance to judge the effects of SRM or CDR. For many, this question has been framed along the lines of Langdon Winner's decades-old provocation: What (existing) political systems might these (new) technical schemes require, or at least be highly compatible with?[7]

Answers to this question have so far largely required the reconciliation of the material affordances of geoengineering to traditional political categories. Szerszynski et al., for example, have made the strong claim that SRM is distinctly incompatible with liberal democracy.[8] They argue (in the pattern of Winner) that the complex commitments to scale and time required by geoengineering would necessitate centralist, autocratic governance, as its technically obfuscated risks and perceived urgencies would not easily accommodate popular dissent.[9] This approach is not unlike that of geoengineering chemtrail conspiracy theorists who, although concerned with an imagined present rather than future, anxiously articulate their mistrust of large-scale technical systems and officially sanctioned knowledge about their management. Democracy in this view is undone by systems that engender and intensify the sociotechnical conditions described in Ulrich Beck's concept of risk society, in which the polis can neither comprehend nor effectively and democratically negotiate the risks that emerge as the unanticipated by-product of industrial modernity.[10] The solution that follows, in turn, is to defend liberal democracy through strident techno-political risk aversion, rejecting the governance constraints of large technical systems.

Other analysts and theorists have claimed the converse to be true. According to Dalby, the scale and complexity of geoengineering instead create "a large technical incentive" for broad and transparent cooperation between states.[11] Like the interconnected material networks of the biosphere within which geoengineering intervenes, the forms of social coordination necessary for the distribution of such actions, if they are to succeed, require broad social license and international buy-in. The democratic quality of individual member states within this alliance is less of an urgent concern to Dalby, as there are (and have always been) a plurality of political ideologies within the terrain of international affairs. This broad operating space, however, invites considerations of the influence of civil societies and social movements that could prove decisive in addressing the particulars of a geoengineering program (such as land reform), forcing environmental justice concerns into discussion and action.[12] There are some encouraging signs to this end emerging from the U.N. Framework Convention on Climate Change (UNFCCC). Through this body, international policy and scientific experts have recognized climate change as a "collective action problem at the global scale," while diplomats, environmentalists, and social movements have parlayed years of organizing into a significant (although still evolving) unilateral agreement inclusive of both geoengineering and climate justice concerns.[13] Christiana Figueres, the former executive secretary of the UNFCCC, praised the achievement as an unparalleled feat of collaborative and noncoercive consensus: "fundamentally a convention of human rights and of peace."[14]

How, then, to reconcile these two dissonant perspectives? At first, there is a matter of imbalanced rhetorical address that must be compensated for if comparisons are to be fairly made. Szerszynski et al. critique only SRM, while the UNFCCC and Dalby have staked their optimism primarily on the reparative potential of CDR. The former position, moreover, is invested in the defense of a traditional model and valuation of liberal democracy and civic transparency (perhaps somewhat inflected by Cold War rhetoric), while the latter camp focuses their assessment on international political coordination and governance.

For the purposes of this chapter, however, it must be noted that there is a significant common analytic between these views that does allow for a joint analysis of current geoengineering discourse. This lies in the ways in which technological development and implementation are conceived. In a manner that perhaps entertains a degree of technological determinism, geoengineering is often jointly imagined as a set of stable technologies that produce or reinforce particular, existing political outcomes and thus should be either pursued through appropriate channels or abandoned as a collective moral hazard. That this same category of technological innovation can be pursued to such dissimilar ends suggests that this approach suffers from an impoverished understanding of what technology is or does when it is embedded in multiscalar, socially tumultuous terrains. Reconciling this impasse and enriching our analyses of SRM and CDR, I argue, require that we contest the monolithic status of geoengineering as technology (with all that word's history in analyses of alienation, disruption, and imposition) and reimagine these practices as forms of infrastructure. Such a move would not serve to dismiss the many material and political concerns of CDR and SRM implementation but would rather redirect these questions into a vocabulary that is better equipped to acknowledge and address them.

Geoengineering as Technology

The first and most significant clarification that must be made in this discussion is the distinction between making and provisioning the climate. The majority of critical scholarship about geoengineering falls into the former frame, taking these technologies as fixed tools to achieve a given end. The philosophical ramifications of making the climate and the forms of political redress within this system of world manufacture are of the foremost concern to these critics. Ibon Galarraga and Bronislaw Szerszynski, for example, have jointly argued that "the discussion of the science, governance, and ethics of geoengineering has been dominated, explicitly or implicitly, by a particular 'imaginary,' a particular idea of what it is to make something: that of 'production.'"[15] The authors identify three such modes of global making—the climate architect, the climate artisan, and the climate artist—drawing from Aristotelian and Heideggerian philosophies of technology. No model is ultimately deemed sufficient, however, as these forms of production either do not apply to complex systems, require Sisyphean commitments to laboring over the climate, or reduce and diminish the world's many poetic potentialities into a singular form of equipment.[16] The result, they claim, is an inadequate theoretical capture of technologies like SRM, leading to potentially disastrous outcomes. An alternative approach, however, is not offered.

Yusoff, similarly, has framed the attendant geopolitical challenges of geo-engineering in terms of the discourses following its unprecedented scales of enactment.[17] Geoengineering, she argues, proposes an affectively laden way of understanding the earth as a "geoengine"—one that "can be altered, modified, and engineered on a global scale . . . open to forms of making."[18] Such an understanding of this singular and titanic earth technology is underscored by her emphasis on the ontological politics of this turn. The earth, revealed in this light, is presumed to be a dynamic object with ontic facticity. Questions of social formation and difference presumably no longer matter in the face of how this new global machine is made and how its making constitutes a new geopolitical world order.[19]

Clive Hamilton, to provide a final example, also constructs his argument around a focus on geoengineering as a distinct techno-philosophical form. His analysis of the debates over geoengineering within scientific and policy circles takes the narrative form of a mythic struggle between the Prometheans, driven by a desire to exert total control and technical mastery over the earth, and the Soterians, motivated foremost by protection and safety and thus unlikely to embrace speculative technologies except in cases of the last resort.[20] Either side, which respectively takes their names from the Greek gods of technology and of preservation, feature in his account as followers of distinct paths for and against geoengineering, focusing on the singular question of whether this particular Pandora's box should be opened. Once this happens, Hamilton presumes, debate will end and a path will be irreversibly chosen. The considerable popular and political faith in technological fixes further frames and diminishes Hamilton's hope in the Soterians' prospects: "Technological thinking structures our consciousness in a thousand subtle ways that make climate engineering attractive, indeed, almost inevitable. Prometheans rule."[21]

However, this emphasis on production and on technology as a hegemonic category and political force, even if well established in scientific and popular accounts of geoengineering, merits further scrutiny. Although these authors do not fully disclose their environmental politics, the terms of their debate imbricate this work within a wider set of values and ontological distinctions that, by many of their own accounts, can no longer be taken as a given. If, as Szerszynski and his coauthors have argued, "there is an urgent need to make explicit the particular way in which SRM [and other forms of geoengineering are] being constituted as a technology,"[22] then a more nuanced analysis of the character of technology in these discussions is all the more required. It is my contention that many of the weaknesses of the approaches to geoengineering so far can be revealed through an alternate approach to its political and technical entanglements. This can be accomplished by a shift in frame to consider geoengineering as infrastructure and, with it, a shift from an emphasis on production toward the reproductive labor of earth systems and their maintenance.

Infrastructural Nature/Cultures

The first theoretical gain from this change would be to unpack and interrogate the implicit moral arguments against geoengineering that stem from the perceived impropriety of crossing geologic scales or blending organic and inorganic systems. As many feminist, Indigenous, and Marxist scholars have forwarded, much of modernity's social order is predicated on the separation of human relations from nonhuman nature—a historic event with considerable consequences for women and preproletarianized peoples.[23] The nature/culture divide is still a significant part of the anthropological schema of much of the Global North, and proposals or actions that violate such distinctions risk the creation of category horrors—hybrid forms that are illegible or irritating to the structural logics of a culture.[24] Such reactions are well evidenced in the anxieties attendant to imaginaries of geoengineering (often registered through aesthetics or affects—a transgression of Eden, a Frankenstein world, or a rude ending to the principles of nature conservation). The question to ask would therefore not be, as Hamilton poses, whether to give in to prior Promethean desires for technical mastery at the expense of an a priori nontechnical nature but rather be how such sociotechnical systems differently entangle already imbricated human and nonhuman worlds and to what effect.

This is a point that has already been well explored within the study of infrastructural formations, a field in which strict divisions between technology and the environment are largely untenable. Ashley Carse, for example, centers watersheds and hydrological systems in his study of the Panama Canal megaproject. In order to move ships through locks, the canal requires massive quantities of fresh water pumped in and out of its system to keep boats afloat during their transit. This watershed in turn depends on the nearby rainforests that must be preserved from further industrial or campesino expansion if the water is to continue to flow. As such, Carse argues that "infrastructures both transform and depend on the ecologies that they cross."[25] It is an error, therefore, to see technical systems apart from the environments that provide much of their foundations. Infrastructures are always already ecological, founded on processes of relationship-building within and beyond the immediate community of their use.[26]

The environment in this configuration can therefore be further theorized as a kind of "infrainfrastructure." It is the material basis and background through which infrastructural systems take form, power themselves, and further distribute that power. Ecosystems, to this end, operate as second-order infrastructural systems. As Carse's case study demonstrates, such systems can also be sociotechnically produced, repaired, and contested as part of the larger stakes of resource provisioning. Looking ahead to the twinned futures of infrastructural and ecological change, he sees the rise of landscape design as an integral part of infrastructural development, perhaps pulling these second-order systems into the foreground of how technical systems are conceived. He warns, however, that the results are likely still to produce social and ecological winners and losers and thus require a politics that addresses the unique imbrication of organic and inorganic provisioning.[27]

Paul Edwards's earlier multiscalar approach to infrastructure supports Carse's analysis. "Nature," Edwards argues, is "in some sense the ultimate infrastructure."[28] This declaration follows from a far-reaching study of infrastructural systems as characterized by their behavior over centuries-long stretches of time. As technical forms that amplify, filter, or reproduce natural forces, Edwards argues that "infrastructures constitute an artificial environment" and thus "simultaneously constitute our experience of the natural environment, as commodity, object of romantic or pastoralist emotions and aesthetic sensibilities, or occasional impediment."[29] As a result, "to construct infrastructures is simultaneously to construct a particular kind of nature, a Nature as Other to society and technology."[30]

In this light, the great scandal of geoengineering is that this long-standing (but by no means inherent) nature/culture dialectic ceases to be possible when nature is so vividly technologically constituted all the way down. Within this shift from second- to first-order infrastructural systems, environments can no longer be accommodated in pastoral frames or geopolitical configurations that imagine them to be inert, constant, or given. This disturbs normative categorical thinking. The challenge, however, is to theorize both organic and engineered systems together in such a way that complicates simple technological determinism and articulates calls for concern and mutual care in a language that extends beyond detached moral longing.[31] This would entail a shift from analyzing geoengineering as the production of a new and hybridized earth system toward an analysis of geoengineering as the social and technical reproduction of environments and environmental relations—a meeting of first- and second-order infrastructures. This would better allow us to keep the human and nonhuman winners and losers of particular proposals in mind rather than mourning the purity of an external nature that is both a social construction and a set of material relations already imperiled by climate change.[32] This, in turn, demands that the language of prior political formations be examined and revised.

Infrastructure's Political Affordances

Traditional categories of politics may certainly adhere to geoengineering, especially when considered as a technological imposition. Troubled national sovereignty, collateral damage, and the inequitable distribution of decision-making power within the emergent configurations of SRM and CDR are all pressing issues that require serious consideration. What I would like to suggest, however, is that infrastructure studies have an existing vocabulary for political claims-making and provisioning that can allow us to better foresee the evolving role climate and climate interventions are currently playing as a subject or medium of collective movements. Such a frame takes sociotechnical systems as politically formed and reformed rather than inert technologies imposed from afar. This vocabulary further invites analyses of coalitions and solidarities that are more geophysical than national or electoral and as such stand to extend our concerns beyond liberal notions of democratic or antidemocratic social orders. Presuming that some form of geoengineering will play a role in future climate responses, it is essential to analyze its potential not only to produce particular outcomes that are compatible or incompatible with the goals of climate justice but also to gather and constitute new polities through infrastructural politics.

Infrastructure and politics have long been closely linked throughout the development and contestation of the technologies and ideologies of modernity. Gaining access to a shared grid, information system, or regional network has often been a joint project of state making and local social formation.[33] More materially and directly than a passport or the franchise, to be provisioned and connected via infrastructure is to be included in a polity, with bidirectional claims of recognition made by both the provisioner and the provisioned. But increasingly, provisioning is no longer the exclusive purview of the state and includes a wider range of public and private, local and global, communities of care (or extraction). These in turn constitute a diverse set of imagined communities and social formations mediated by the infrastructure that connects them.[34]

Geoengineering and climate change further complicate this trajectory of resource distribution, social mobilizations, and claims-making. In their scale and material impacts, they foster the potential for different kinds of community and citizenship, signs of which can be sighted in recent collective actions against climate change. Contrary to traditional models of geopolitics, in which nation-states autonomously pursue sovereign policies, with their heterogeneous populations perhaps guiding governments through electoral democracy, much of recent actions on climate change have taken the form of unique legal initiatives and social configurations that make claims on geophysical systems and conditions. Guidelines from the Intergovernmental Panel on Climate Change (IPCC), for example, advocate that geoengineering be treated in policy as a collective, international public good, while the organization's executive secretary openly speaks of climate change as an appropriate area for human rights claims.[35] In such openings, new territorial and climatic polities are being formed.

These developments borrow from and innovate on formal and informal models of accessing and demanding services through infrastructure. Whereas David Harvey articulates a "right to the city" through popular claims to resources and capacities for the communal transformation of urban spaces, demands to secure a more stable and traditional geophysical environment might be said to contain the nascent formations of a parallel "right to safe climate."[36] The emergence of such claims subtly anticipates and furthers geoengineering proposals. On the level of international alliances, the 2015 Paris Climate Conference (COP 21) saw the formation of novel coalitions across the Global North and South, most notably the High Ambition Coalition, spearheaded by the governments of the Marshall Islands and St. Lucia and recruiting more than one hundred additional countries, eventually including the United States. This block was largely responsible for the inclusion of the 1.5°C target in the Paris Agreement and, as a result, the commissioning of a forthcoming report on the policies and technical systems necessary to meet this target (of which geoengineering is expected to play a substantial role). Regionally, there have been similar coalitions making claims to climate rights, such as Indigenous circumpolar communities' organizing efforts around "the right to be cold."[37] There have also been a series of legal cases coordinated by Our Children's Trust to bring charges against multiple governments on the basis of generational discrimination via climate degradation. As these demands can increasingly only be met by both significant mitigation actions in the present and reparative efforts toward the climate system in the near future, geoengineering stands to be an increasingly consequential mode of addressing these human rights claims and meeting the demands of these emerging climate constituencies, even under ambitious carbon mitigation scenarios.[38]

This suggests that the question of geoengineering's political tendencies might be best approached not through an understanding of liberal democracy but through the kind of democracy described by Timothy Mitchell as "a battle over the distribution of issues, attempting to establish as matters of public concern questions that others claim as private . . . as belonging to nature . . . or as ruled by laws of the market."[39] New generationally, territorially, and thermally defined interest groups, constituted beyond the nation-state, have made notable progress to this end. Making claims on climate—demanding climate provisioning—is the organizing principle that constitutes these nontraditional polities. The qualities of the system of governance to which these claims are addressed (whether they are "democratic" in a classical sense) seem globally much less consequential than the new forms of solidarity, interest, and rights being articulated through contestation.

To summarize, when analyzed as infrastructure, geoengineering can no longer be taken as a simple technofix—such matters are not simple, because new infrastructures imply new social formations that shape and are shaped by them. These emerging social categories do not exclude the historic weight or future importance of other forms of identity politics, nor is their political character inherently progressive or determined in advance. This process of climatological claims-making and group constitution, however, remains an important consideration for any theorization of speculative geologic politics or a Gramscian war of position over a livable earth for all.

However, just as the history of infrastructure describes the conditions through which claims-making constitutes new polities, it also reveals the extent to which large sociotechnical systems can be constructed in ways that structurally inhibit social solidarities. When market inequalities are deepened, when services are selectively or unequally provided, and when fears of security or greater incentives toward comfort are prioritized above considerations of the public good, infrastructure's potential for strong community formations can be overturned. There is justifiable concern around the potential for carbon trading and patent races to favor the marketization of geoengineering above and beyond its potential to approach the global climate as a commons.[40] Similarly, there is the pressing question of how to fairly manage and distribute secondary effects of SRM and CDR, ensuring that the regulation and reconstruction of climate norms that do not disintegrate into secessionary regional climate enclaves, enacting a form of eco-apartheid at unprecedented scales.[41] Like Stephen Graham and Simon Marvin's classic analytic framework of splintering urbanism, the emerging geologic politics of climate change could lead to either expansive solidarities or their exclusions.[42] This is a more socially complicated, and more politically vital, framework for analysis than that of technological solutionism—for or against. It demands engagement with evolving scientific and social complexity and rejects a politics of determinism.

Infrastructure as Care

A final theoretical gain that follows from the analytic shift from technology to infrastructure lies in the greater vocabulary infrastructure studies have to offer when attending to the ethics of care required by geoengineering's impactful undertakings. This discussion has not taken center stage in current geoengineering policy or ethics discourse, as it depends on the prior framing of the climate as both a collective undertaking and public good.[43] With these conditions secured, it is then possible to have a more nuanced discussion of the timescales, labor, and commitments of geoengineering, not as a quick fix but as a difficult and durational effort of repair.

First and foremost, it is again important to consider the ways in which a technocentric approach may overemphasize Western humanist values ill-suited to the scale and stakes of geoengineering. Galarraga and Szerszynski, for example, are invested in probing the ways in which speculative forms of climate making would correspond to deleterious forms of human making in turn. The project of geoengineering will be unending, they insist, requiring continuous forms of work, study, and adjustment that must be maintained over the course of centuries. As a result, "the climate artisan would thus be not homo faber, the human being as the fabricator of the enduring things of a made world, but animal laborans, the laboring animal who serves the endless processes of life's self-maintenance."[44] James Lovelock has adopted a similar position, arguing that climate engineering would result in nations "enslaved in a Kafka-like world in which there is no escape."[45] Undertaking such a project, these scholars argue, would be costly and dehumanizing.

Infrastructure studies, conversely, approach the work of maintenance as a necessary component of any social fabric and a much-needed sobering measure in the face of apocalyptic fears. When assessing new sociotechnical systems, parallels can be drawn to existing forms of planning and labor that go into supporting contemporary electrical grids, global logistics, and communications networks. These lines of continuity challenge the ways in which geoengineering's durational labor requirements are framed as an exceptional case. This view of maintenance as quotidian, ongoing, and valuable work further challenges the tacit gender and class bias in the forms of technical knowledge and work that are considered to be ennobling compared to those that are denigrated within the history of science and the wider philosophy of technology fields. Lovelock, Galarraga, and Szerszynski, in reducing such labor to subhuman status, perpetuate a social order that must be challenged from both a social justice perspective as well as a more utilitarian analysis of the labor formations that are needed to meet the challenges of a changing climate.[46] CDR, with its durational commitments to undoing the cumulative effects of fossil fuel use, could embody these ethics of repair in particular.[47] Geoengineering would certainly require a significant amount of study, investment, and maintenance, but these forms of work can be seen as the expression of collective social values rather than a collectively degrading ball and chain.

An openness to labor and repair as productive analytics allows for clearer demands to be articulated about geoengineering's labor standards—if work itself is not the principal objection, then we can begin to nuance what standards of work, and for whom, are worth fighting for. Framed as reparative infrastructure, CDR in particular becomes thinkable within the blue/green coalitional work of energy democracy or in calls for an expansive Green New Deal.[48] The potential for negative outcomes, similarly, is clearer in sited analyses of infrastructural work, dispossessions, and enclosures. Infrastructure is by no means inherently progressive, and this only serves to emphasize the need to frame its risks and benefits as questions of political contestation, not as fixed or inevitable outcomes of a technological pathway.

Conclusion

In this chapter, I have traced common approaches to the politics of geoengineering, outlined some of their shortcomings, and suggested the utility of a new approach based on the analytic possibilities of infrastructure studies. As the Paris Agreement unfolds and the technical backdrop to its targets comes forward with greater urgency, I maintain that fuller discussions of geoengineering's potential will require a reassessment of what we take to be technology, and to what end. Looking at geoengineering as an infrastructural formation for the provisioning of climate has three principal advantages: it allows divisions between the artificial and the natural to be productively rethought as continuous categories rather than opposites, avoiding unhelpful moral judgments; it provides an existing vocabulary for claims-making and social contestation that can be more easily adapted to the nontraditional political configurations of geophysical social formations; and it suggests ways in which the labor and durational commitments of geoengineering could express an ethics of care and reparations rather than dystopian obligations to toil.

Geoengineering, without a doubt, constitutes and is formed from significant disruptions to environmental norms and governance systems, with repercussions that will shake geopolitical structures and forms of social belonging alike. In this space of disruption, however, there is the potential to remake the world and its politics in a more just fashion. Understanding geoengineering's potential and better assessing its accompanying hazards, I argue, is best accommodated in the shift from a philosophy of technology approach to that of infrastructure studies. In a > 400 ppm (parts per million) CO_2 world, we must respond to the growing claims made in the name of a right to a safe climate. As our collective carbon budget rapidly runs out, scholarship attuned to recognizing and uplifting such claims is all the more necessary.

Notes

1 Andreas Malm, *Fossil Capital: The Rise of Steam Power and the Roots of Global Warming* (London: Verso, 2017); Timothy Mitchell, *Carbon Democracy: Political Power in the Age of Oil* (London: Verso, 2011).
2 Myles Lennon, "Decolonizing Energy: Black Lives Matter and Technoscientific Expertise amid Solar Transitions," *Energy Research & Social Science* 30 (August 1, 2017): 18–27, http://doi.org/10.1016/j.erss.2017.06.002.

3   Kathryn Yusoff, "The Geoengine: Geoengineering and the Geopolitics of Planetary
    Modification," *Environment and Planning A* 45, no. 12 (December 1, 2013): 2800, http://
    doi.org/10.1068/a45645.

4   Whose history, of course, remains an important political question, especially as it pertains
    to the question of reparations.

5   Simon Dalby, "Geoengineering: The Next Era of Geopolitics?," *Geography Compass* 9,
    no. 4 (April 1, 2015): 191; Nigel Clark, "Geoengineering and Geologic Politics," *Environ-
    ment and Planning A* 45, no. 12 (December 1, 2013): 2827.

6   Yusoff, "Geoengine," 2801.

7   Langdon Winner, "Do Artifacts Have Politics?," *Daedalus* 109 (1980): 130; see also
    Markusson et al., this volume, chap. 14.

8   Bronislaw Szerszynski et al., "Why Solar Radiation Management Geoengineering and
    Democracy Won't Mix," *Environment and Planning A* 45, no. 12 (December 1, 2013): 2811.

9   Szerszynski et al., 2812. See also Malm, this volume, chap. 10.

10 Ulrich Beck, *Risk Society: Towards a New Modernity*, trans. Mark Ritter (London: Sage,
    1992), 260.

11 Dalby, "Geoengineering," 197.

12 Dalby, 197; see also Schneider and Fuhr, this volume, chap. 4.

13 Intergovernmental Panel on Climate Change (IPCC), *Fifth Assessment Synthesis Report*
    (Geneva, Switzerland: IPCC, 2014), 76. Its latest achievement, the Paris Agreement, is of
    course woefully insufficient in its promised decarbonization commitments and the matter
    of North-South climate reparations. Accordingly, while this work remains an important
    achievement for a climate politics focused on issues of representation and consensus, its
    material gains remain highly disappointing from an anticolonial climate perspective.

14 Christiana Figueres, "Human Rights and Climate Change" (speech, New York University,
    January 27, 2016).

15 Maialen Galarraga and Bronislaw Szerszynski, "Making Climates: Solar Radiation
    Management and the Ethics of Fabrication," in *Engineering the Climate: The Ethics of
    Solar Radiation Management*, ed. Christopher Preston (Lanham, Md.: Lexington Books,
    2012), 20.

16 Galarraga and Szerszynski, "Making Climates," 12–19.

17 Yusoff, "Geoengine," 2805.

18 Yusoff, 2806.

19 Yusoff, 2806.

20 Clive Hamilton, *Earthmasters: The Dawn of the Age of Climate Engineering* (New Haven,
    Conn.: Yale University Press, 2013), 18.

21 Hamilton, 209.

22 Szerszynski et al., "Why Solar Radiation Management," 2810.

23 See, for example, the work of Silvia Federici, Maria Mies, or Zoe Todd.

24 Carolyn Merchant, *The Death of Nature: Women, Ecology, and the Scientific Revolution*
    (San Francisco: Harper & Row, 1980).

25 Ashley Carse, *Beyond the Big Ditch: Politics, Ecology, and Infrastructure at the Panama
    Canal* (Cambridge, Mass.: MIT Press, 2014).

26 Carse, 219; see also Nicole Starosielski, *The Undersea Network* (Durham, N.C.: Duke
    University Press, 2015).

27 Carse, *Beyond the Big Ditch*, 221.

28 Paul N. Edwards, "Infrastructure and Modernity: Force, Time, and Social Organization in
    the History of Socio-technical Systems," in *Modernity and Technology*, ed. T. Misa, P. Brey,
    and A. Feenberg (Cambridge, Mass.: MIT Press, 2004), 196.

29 Edwards, "Infrastructure and Modernity," 189.

30 Edwards, 189.

31 For a discussion on "matters of care" versus "matters of concern" in technoscience, see María Puig de la Bellacasa, *Matters of Care: Speculative Ethics in More than Human Worlds* (Minneapolis: University of Minnesota Press, 2017), 42.

32 For more on the fraught politics of purity ethics, see Alexis Shotwell, *Against Purity: Living Ethically in Compromised Times* (Minneapolis: University of Minneapolis Press, 2016).

33 Stephen Graham and Simon Marvin, *Splintering Urbanism: Networked Infrastructures, Technological Mobilities and the Urban Condition* (London: Routledge, 2001).

34 Infrastructure, to this end, follows a similar pattern of group self-recognition described in Benedict Anderson's *Imagined Communities* (London: Verso, 2006).

35 Ottmar Edenhofer et al., eds., *IPCC Expert Meeting on Geoengineering* (Geneva, Switzerland: IPCC, 2011), 35; Figueres, "Human Rights and Climate Change."

36 David Harvey, "The Right to the City," *New Left Review*, no. 53 (2008): 23.

37 Sheila Watt-Cloutier, *The Right to Be Cold: One Woman's Fight to Protect the Arctic and Save the Planet from Climate Change* (Minneapolis: University of Minnesota Press, 2018).

38 See, for example, James Hansen et al., "Young People's Burden: Requirement of Negative CO_2 Emissions," *Earth System Dynamics Discussions*, October 4, 2016, 1–40, http://doi .org/10.5194/esd-2016-42.

39 Mitchell, *Carbon Democracy*, 9.

40 Yusoff, "Geoengine," 2800; Melissa Leach, James Fairhead, and James Fraser, "Green Grabs and Biochar: Revaluing African Soils and Farming in the New Carbon Economy," *Journal of Peasant Studies* 39, no. 2 (April 1, 2012): 42, http://doi.org/10.1080/03066150 .2012.658042. See also Sapinski, this volume, chap. 1; and Parenti, this volume, chap. 9.

41 Daniel Aldana Cohen, "New York Mag's Climate Disaster Porn Gets It Painfully Wrong," *Jacobin*, July 10, 2017, http://www.jacobinmag.com/2017/07/climate-change-new-york -magazine-response.

42 Graham and Marvin, *Splintering Urbanism*, 228.

43 For earlier considerations of this perspective and the constraints that threaten it, see in Holly Jean Buck, "Rapid Scale-Up of Negative Emissions Technologies: Social Barriers and Social Implications," *Climatic Change* 139, no. 2 (November 1, 2016): 165; and Holly Jean Buck, Andrea R. Gammon, and Christopher J. Preston, "Gender and Geoengineering," *Hypatia* 29, no. 3 (June 1, 2014): 661.

44 Galarraga and Szerszynski, "Making Climates," 231.

45 James Lovelock, "A Geophysiologist's Thoughts on Geoengineering," *Philosophical Transactions of the Royal Society A* 366 (2008): 3888.

46 Anna Lowenhaupt Tsing, "A Feminist Approach to the Anthropocene: Earth Stalked by Man" (lecture, Helen Pond McIntyre '48, Barnard Center for Research on Women, Barnard College, New York, November 10, 2015), https://bcrw.barnard.edu/videos/anna -lowenhaupt-tsing-a-feminist-approach-to-the-anthropocene-earth-stalked-by-man/.

47 See, for instance, the argument for CDR as remediation in Hansen et al., "Young People's Burden."

48 See Holly Jean Buck, "A Best-Case Scenario for Putting Carbon Back Underground," *Science for the People*, July 24, 2018, http://magazine.scienceforthepeople.org/best-case -scenario-carbon-underground/, for an example of this potential, articulated through an infrastructural frame. A larger issue, unaddressed in this chapter, concerns the character and means of securing such a compromise between labor and capital and how such terms might hold at the global scale of this problem. See Malm, this volume, chap. 10 for further thoughts to this end.

Part IV

Geoengineering

A Class Project in the Face
of Systemic Crisis?

12

Geoengineering and Imperialism

RICHARD YORK

The age in which we live, referred to with growing frequency by scientists as the Anthropocene, is one defined by global environmental crises. Among other things, human activity has transformed ecosystems around the world, dramatically accelerated the rate of biodiversity loss, altered planetary biogeochemical cycles, and changed the global climate. A number of novel environmental problems have emerged in recent decades, such as radioactive contamination from nuclear weapons and power plants and the dispersal of synthetic toxins and plastics around the world, marking current environmental crises as unique in human history. Of course, many of the hallmarks of our current precarious state, such as deforestation and soil degradation, have been ongoing for millennia but have recently greatly expanded in scale. Nonetheless, we are regularly presented with the promise—by corporate managers, politicians, economists, and engineers among others—that technological innovations and the marvels of the market will overcome any challenges we face.

It seems that modernity has always been poised to solve the environmental problems it creates but never quite solves them. Apologists for the modernization project assert that the only way to solve the problems of modernization is to have further modernization—a new and improved "ecological modernization" that will overcome environmental problems and ecological limits through technological refinements.[1] Part of the arrogance of the modernist mind is its sneer at the past as inefficient and ineffective while it joyously anticipates the clean and green future it presumes is on the horizon. However, as we march toward this brighter, hypermodern future, it is ever receding like the end of a rainbow—as a wide array of evidence shows that modernization processes continue to drive environmental problems around the world.[2] Here I shall argue that since its emergence, capitalism's machinations continually produce environmental and social crises, and those in power continually promise that the continuation of the capitalist project will resolve these crises. However, the expansion of capitalism leads to newer, more global crises. Indeed, the "solutions" to social and environmental problems promised by capital are ways of sustaining capitalism, not sustaining the ecological integrity of the planet or the well-being of humans.

Proposals for geoengineering projects that have become popular in recent years—such as those aimed at addressing global warming by injecting particles into the atmosphere to increase its albedo and thereby reduce solar input[3]—although new in their particulars, are prime examples of the same old capitalist tactic of aiming to solve the problems capitalism itself creates in ways that maintain capitalism while failing to address the fundamental political-economic causes of the problems, such as the drive for profits, accumulation, and endless growth.[4] As a point of comparison to the current moment, I examine an earlier geoengineering project that originated in the interwar years: the Atlantropa project, which aimed to reengineer the Mediterranean system and the rivers of central Africa. This comparison shows how the geoengineering proposals of the twenty-first century fit in a tradition of ill-conceived imperialist projects that treat the earth and humanity as things to be manipulated to sustain prevailing power relations. At its heart, geoengineering is an imperialist project. Consistent with previous waves of imperialism, it seeks to solve the environmental problems created in the global core of "modernized" nations by controlling resources, biophysical systems, and the lives of others around the world. The key lesson from my assessment is that our environmental and social problems originate in the political-economic order and the structure of modernity and that addressing them requires challenging this order, not hoping for salvation from engineers, technology, and the free market.

Capitalism's Crises and Solutions

European colonialist projects began to take on their ferocious form in the fifteenth century, driven by greed and a hunger for resources. The desire to control trade routes and access natural resources from other continents was a central concern from the start. It was in the context of the aggressive expansion of European power around the world that capitalism was born. And in turn, the development of capitalism came to drive further global expansion of European powers and to intensify colonialism.

Ecological crises always have been a central part of the capitalist-colonial project. Capitalism continually generates environmental crises due to its rapacious consumption of resources to generate profits for corporations and its instrumental approach to the environment,[5] and one of its "solutions" to the crises it creates (e.g., soil depletion, forest loss, and the undermining of ecosystems) has been to expand its imperial reach so as to find new resources to replace the ones it has destroyed closer to the European core.[6] In addition to colonization to gain access to new resources, the other standard approach of capital to addressing environmental crises is to develop new technologies—such as responding to the loss of availability of fuel-wood in Europe due to deforestation by increasing the exploitation of coal—and thereby creating new environmental crises as part of the attempt to overcome old ones.[7] Clark and York refer to the process as one of "rifts and shifts," where capitalism creates an ecological rift—that is, disrupts a particular ecological process—and seeks to address the problem with either/both a shift to exploitation of the same resource in a new location and/or to a new type of technology or resource (which also often involves exploiting/ colonizing a new region of the globe).[8]

One prime example of this process of rifts and shifts is how European powers handled the soil crisis. As Marx recognized, the emergence of capitalist agriculture created a rift in the soil nutrient cycle by preventing nutrients from returning to the soil.[9] This occurred because capitalism pushed to expand agricultural production through mechanization as it increasingly moved agricultural products to newly growing industrial centers. Thus rather than food and fiber being consumed locally, as they typically had been, they were moved away from the countryside, and therefore, nutrient-laden waste (human, animal, and agricultural) instead of being recycled into the soil became an urban pollution problem, contaminating waterways and accumulating on city streets and in landfills. One of the "solutions" to the degradation of soil fertility that was developed by capitalists was to scour the world for bird guano, which is rich in nutrients, to replenish European soils.[10] This led to imperial grabs and even wars to seize control of small oceanic islands, such as those off the west coast of South America, where bird colonies had long been established and guano had accumulated in layers for ages. Imperial powers often imported forced labor to mine the guano and quickly ravaged islands across the globe.[11] This guano imperialism disrupted bird colonies and harmed fisheries, since the bird colonies and guano islands, before being scraped clean, served as sources of nutrients for marine ecosystems. Thus the soil rift was addressed by a shift to guano as fertilizer, which led to a new suite of environmental problems and social injustices. However, guano could still not meet the needs of agriculture, and in due course, there was yet another shift—this time to chemical fertilizer following the development of the Haber-Bosch process of artificial nitrogen fixation in the early twentieth century. This shift led to a new suite of environmental problems as chemical fertilizer came to be the dominant force in the global nitrogen cycle. The production of chemical fertilizer is energy intensive and fossil fuel dependent, and the application of fertilizer greatly increased eutrophication of water systems.[12] In fact, through its recurring process of rifts and shifts, the colonial project disrupted agricultural practices and soil fertility regimes around the globe, leading to soil degradation and famine.[13]

Capitalism's approach to the crisis of soil fertility is a model for how it handles environmental crises in general. Similar dynamics played out with regard to energy resources. Industrialization's voracious appetite for energy quickly depleted forests in Europe as wood was consumed for fuel. This rift was addressed with a shift to fossil fuels (although the destruction of forests continued and wood, where available, continued to be burned as fuel). The rise of fossil fuels, of course, created a new series of rifts, anthropogenic climate change most notable among them. In fact, climate change was created by various shifts, including changes in agricultural practices, since soils store large amounts of carbon. Now as climate change—perhaps the most threatening rift of our era[14]—is beginning to run out of control, a new series of proposed shifts has come to the fore. Not only does capitalism promise us new energy sources, like nuclear power (which creates new types of ecological rifts), but it promises us a grand solution to climate change: geoengineering.

To put it in context, geoengineering is just the latest in a string of capitalist solutions to the problems capitalism creates. Importantly, as with earlier "solutions" to environmental crises, the focus on geoengineering tends to lead to framing the problem as merely a technical one, not one that stems from the structure of capitalism or the nature of the colonial project. Geoengineering threatens to create new problems, many of which are likely unanticipated, while also promising to maintain the political-economic status quo and allow capitalists to continue to accumulate more capital. Thus proposals to engineer the climate[15] are both new, in that they propose novel technological solutions, and old, in that they fit with the historical pattern of rifts and shifts based on technological fixes that obfuscate the root causes of environmental crisis in capitalism. Although the specifics of contemporary climate engineering proposals are new, the urge to engineer the planet has been part of the modernization project for some time. Early in the twentieth century, techno-optimists proposed an ambitious engineering project of extraordinary scale, the Atlantropa project, as a solution to some of the problems then facing European powers. I now turn to this project, since it provides insight into the geoengineering projects proposed in the twenty-first century.

The Atlantropa Project

In the interwar years, Europe faced many of the types of problems the world faces today, ecological crises and social strife among them. Energy was in high demand to power the industries on which European economies had come to depend, and there were growing concerns across nations about access to energy resources. The ever-growing demand for food production, timber, and other resources—which were hard to meet due to the depletion of soil, deforestation, and other forms of land degradation caused by modernization—left many Europeans desiring new territory to settle and often looking across national borders for land to take. This demand for energy and land was connected with social unrest, as many people were left in poverty. The European and global economies were in the midst of a severe depression. These problems were connected to the jarring modernization driven by capitalism that had been sweeping Europe for decades, leading to war and revolution. In the eyes of many, Europe had serious problems—insufficient land and energy, economic depression, and civil unrest—which called for a solution.

Recognizing these challenges and with a desire to provide energy and land to help unify Europe and build an enduring peace in the aftermath of the Great War, in the late 1920s, the German architect Hermann Sörgel began to develop a plan of enormous scale and ambition to engineer the hydrology of the Mediterranean Sea Basin and the rivers of central Africa.[16] His extraordinary plan—which he developed and promoted throughout the 1930s and which was given consideration in government, business, and public spheres until the early 1950s—centered on building a huge dam to block the Straits of Gibraltar, cutting off the Mediterranean Sea from the Atlantic Ocean and connecting Europe and Africa into a new supercontinent: Atlantropa.

Sörgel's plan was based on the recognition that the evaporation rate of the Mediterranean Sea was greater than the input of water from rain and tributary rivers, so its water level would fall once it was cut off from the Atlantic.[17] This would open up new land as the sea level fell and provide the potential for the generation of a huge amount of electricity through hydropower. Sörgel calculated that the drop from the Atlantic to the Mediterranean would reach one hundred meters in little more than one hundred years. Thus he planned to build a hydroelectric plant at the Gibraltar dam, which could take advantage of the effectively inexhaustible reservoir of the Atlantic. He also proposed to later build a dam between Tunis and Sicily to lower the eastern Mediterranean by another one hundred meters. He further proposed building dams on the Suez Canal, the Dardanelles, and several major rivers. Together, all these dams would provide Europe with a vast quantity of hydroelectricity.

Sörgel further developed his plan to include creating an inland sea in the Congo Basin by damming the Congo River. Additionally, he proposed diverting part of the Congo's flow northward to create an artificial lake system in the Sahara.[18] He believed that these new inland bodies of water would help make the African climate milder and therefore more amenable to European settlement. Thus Sörgel was aiming to engineer the climate, the hydrological system across three continents (since Asia would be affected too), and the dry land area accessible to Europeans.

Despite, or perhaps because of, the unprecedented scope and scale of the proposed project and the many technical challenges and unknown consequences it entailed, Atlantropa caught public attention, being featured in news reports and publications, and was taken seriously by political and business leaders, particularly in Germany.[19] The context of the Great Depression provided fertile ground for such a proposal. The massive engineering project promised to help spur economic growth and industrial development by providing electricity and to curb unemployment by generating jobs. It also offered new land area, which Sörgel hoped could ease political tensions among European powers. With the rise of Nazism and its demand for *lebensraum*, the proposal fit with many concerns of the time (although Hitler had set his sights on land to the east, so he was not enticed by the proposal).

The fundamental Eurocentrism of the proposal had multiple dimensions. Despite its noble objective of building a lasting peace by encouraging cooperation among European nations, it was entwined with the ugliness of colonialism and racism. Sörgel's thinking was grounded in a desire to maintain and enhance the status of Europe in the world. His view of Africa as a source of resources and land for European settlers was unambiguously racist and imperialist and showed his concern for a just peace did not extend beyond European borders. Additionally, Sörgel wished to strengthen Europe so that it could stand up to what he saw as the "Yellow Menace" to the east.[20] He also feared the rising power of the United States, thinking that Europe could only maintain its status in the world if it were unified politically and economically as well as fortified with more energy and land. Sörgel's thinking was widely shared in Germany and other European nations both before and after the Second World War.

The Atlantropa project is in many ways emblematic of modernity and fits with a drive for modernization that was not restricted to capitalist or Western nations alone. The Soviet Union was also enthralled with macroengineering projects. In the era before the space race, there was a hydro-technical competition among nations that was established by the 1930s but later intensified with the emergence of the Cold War. The Soviets admired the Atlantropa project and developed the Davidov plan for Siberia. The Davidov plan called for a sweeping project of damming, irrigation, and hydroelectric development that would divert numerous Siberian rivers through the Aral Sea into the Caspian and dam the Ob and Irtysh rivers, creating a quarter-million square kilometer lake and inundating huge areas of tundra with the hopes of moderating the climate and increasing precipitation. Just as greedy Europeans looked at Africa as a land for the taking, Russian leaders looked at Siberia and its peoples as theirs to do with as they wished. Thus imperialist and racist ambitions were not only rampant in the West or in capitalist nations but part of the modernist craze for controlling nature and the lives of Indigenous people around the world.[21]

The reasons for the decline of interest in the Atlantropa project are telling with respect to the role it served in maintaining the political-economic status quo. As Gall notes, it was not until "new dreams of atomic energy finally rendered it obsolete in the 1950s"[22] that the Atlantropa project fell out of favor. Thus it was only needed until a new grand technological fix to the world's problems came along. Atomic energy, of course, offered the false promise of energy too cheap to meter and is a failure in that it does not provide a large share of global energy, is not cheap, and presents major risks and problems. Nonetheless, like the Atlantropa project, the promise of atomic energy helped maintain the progrowth ideology that is behind corporate power and nationalist politics.

The Imperialism of Climate Engineering

The central commonality between the Atlantropa project and the latest proposals for climate engineering is that they are representative of the way modern capitalist societies frequently aim to address the problems that modernization has created with more modernization, not with efforts to transform societies so that they are just and sustainable. Additionally, these projects reflect a high-handed and imperialist approach to addressing problems. The Atlantropa project was clearly imperialist in nature, where European powers were looking at Africa as a continent to be seized and controlled. After all, none of the discussions of Atlantropa involved consulting the peoples of Africa about their preferences.

Similarly, proposals to engineer the climate come from affluent nations that are primarily responsible for generating global climate change, and they promote imposing a regime on the world that suits their preferences, not one that addresses inequalities and injustices or that leads to an ecologically sensible economy. Some geoengineering schemes are more clearly imperialist than others, since they have designs on the land in nations outside the capitalist core. For example, well-known physicist Freeman Dyson suggests replacing one-quarter of the world's forests, many of which are in tropical countries, with genetically engineered carbon-eating trees.[23] Such a proposal requires controlling land in many nations and would undoubtedly affect people and ecosystems around the world. A similar scheme that has been growing in popularity among policy makers and modelers for the Intergovernmental Panel on Climate Change (IPCC) is bioenergy with carbon capture and storage (BECCS).[24] BECCS is a purported route to "negative emissions" in that it proposes generating electricity from burning biomass (which removes carbon from the atmosphere when it grows) and pumping the emissions underground for geologic storage. To make a significant dent in the carbon concentration in the atmosphere would require vast, continent-sized plantations managed by agribusiness to produce the biomass.[25] Thus inherently, proposals for large-scale BECCS projects require controlling land in many nations around the world, harking back to earlier colonial projects. Of course, this control likely would not be done in the classical colonial sense of military seizure of the lands involved but would be done through neocolonial political and market mechanisms.

Other geoengineering proposals, such as those aimed at solar radiation management (SRM) via stratospheric aerosol injection and cloud brightening, although not aimed at directly controlling land, can be understood as based on a form of neoimperialism if we consider the atmosphere, not only land, as a space that is being controlled. After all, the atmosphere is as much a part of the environment as land and water, and in our world, there are power struggles over its control. Thus SRM schemes in effect seize control of the atmosphere, and these plans are typically driven by people in affluent colonial nations. Of course, climate imperialism has been going on for some while now, as industrialized nations have emitted huge quantities of greenhouse gases (GHGs) over the past two centuries, which affect the climate and lives of people around the world without their consent. So the latest proposals continue on this imperialist path, presuming the right of industrialized nations to maintain their control of the atmosphere.

Despite the threat of imperialism inherent in the geoengineering project, it is important to recognize that one of the key aspects of technological fixes is not singularly the effects they have when applied. Indeed, like the Atlantropa project, many proposed technological schemes are never applied. The Atlantropa project was discussed for over twenty years but never moved beyond the planning phase. It is entirely possible that large-scale attempts to engineer our climate, like those proposed by Keith,[26] may never occur. However, the presence of these types of proposals serves to distract from focusing on the political-economic causes of social and environmental problems by promising solutions that do not require radical societal changes. Thus in our era, like in the past, the primary function for capital of promoting geoengineering is the promise of solving the climate crisis while allowing the continuation of capitalism and consumption of fossil fuels. The promise of geoengineering can undermine efforts to address the fundamental sources of our environmental crises and prevent societies from transitioning away from fossil fuels and developing more equitable and sustainable social forms.

It is important to note that I am not suggesting geoengineering is part of a conspiracy or that many of the supporters of geoengineering projects are insincere. Indeed, I fully accept that many proponents of climate engineering, such as Keith,[27] take environmental problems very seriously and are looking for ways to address them, just as Sörgel genuinely cared about peace and believed his project would improve the lot of humanity (although his narrow, Eurocentric view of humanity). At issue is not the motivation of any one geoengineering proponent but rather the reasons geoengineering proposals gain attention in political and business circles and are promoted to the general public. People in power seek to address problems in ways that do not undermine their power, and geoengineering, therefore, appeals to global business and political leaders, since it promises to spare the world from a climate catastrophe without requiring social change.

Conclusion

There is little doubt that anthropogenic forces have dramatically altered many aspects of the biosphere, perhaps most notably the global climate. The continuation of these processes, including the rising concentration of GHGs in the atmosphere primarily due to fossil fuel production and consumption and the alteration of land cover (e.g., deforestation), over the course of the twenty-first century threatens to undermine the viability of human civilizations and produce a mass extinction comparable to those in the geologic record.[28] Although there is widespread denial among politicians and the general public, especially in the United States, about these verities, the scientific community and many other people, including some prominent political and business leaders, acknowledge the seriousness of global climate change and other environmental problems. However, the "solutions" to these problems—the "ecological modernization" of production processes and geoengineering projects most prominently—that are proposed by the global elite typically show a misunderstanding of the political-economic forces that have given rise to the global environmental crisis and conceptualize the crisis as principally a technical challenge that can be addressed through high-handed technological projects that allow for the maintenance of the capitalist and imperialist status quo.

Capitalist modernity is born from European imperialism, having its origin in the colonial projects that began in the fifteenth century, when European powers seized control of trade networks, human labor, land, and natural resources around the world. The capitalist-imperialist worldview was based on the presumption that the colonized and enslaved existed to serve the rulers in Europe and, as science and technology advanced through the Enlightenment, that humanity can control and remake nature. The economic engineering of the world and the industrial revolution that the colonialist project initiated led to an array of environmental crises, such as soil depletion and deforestation in many regions. However, the imperial powers never accepted these as emerging from fundamental contradictions of capitalism and sought their solution in technological fixes (e.g., the use of chemical fertilizers and the expansion of coal as a fuel source) and geographic displacements (e.g., the importation into the imperial core of resources from regions all around the world).

Seen in this historical light, modern proposals for geoengineering the atmosphere to address anthropogenic climate change, although new in their particulars, are merely a continuation of the rolling crises and failed and unjust solutions—the rifts and shifts of capitalism—that have characterized the colonialist era. Core powers wreak havoc on the earth by engineering it (without consent from other people or, of course, the earth itself) and then seek to fix these problems by more engineering, decided on by those in power, not by consensus of the masses. The Atlantropa project, as explained here, is merely one historical example that exemplifies this Promethean, Eurocentric arrogance, where all the world, including its peoples, is seen as something to be manipulated so as to maintain the global order for the benefit of those in power. Thus proposals for geoengineering projects are a sleight of hand that takes the focus off the political-economic order and the injustices and ecological crises that are inherent in its operation. If we are to truly address the environmental crises we face, we must change our political and economic systems, not continue the geoengineering and imperialism that created these crises in the first place.

Notes

1   Arthur P. J. Mol, *The Refinement of Production: Ecological Modernization Theory and the Chemical Industry* (Utrecht, Netherlands: Van Arkel, 1995).
2   Richard York, Eugene A. Rosa, and Thomas Dietz, "Ecological Modernization Theory: Theoretical and Empirical Challenges," in *The International Handbook of Environmental Sociology*, 2nd ed., ed. M. Redclift and G. Woodgate (Cheltenham, U.K.: Edward Elgar, 2010), 77–90; Andreas Malm, *The Progress of This Storm: Nature and Society in a Warming World* (London: Verso, 2018).
3   Noah Byron Bonnheim, "History of Climate Engineering," *WIREs Climate Change* 1 (2010): 891–897; David Keith, *A Case for Climate Engineering* (Cambridge, Mass.: MIT Press, 2013).
4   Malm, *Progress of This Storm*; J. P. Sapinski, "Climate Capitalism and the Global Corporate Elite Network," *Environmental Sociology* 1, no. 4 (2015): 268–279.
5   Andreas Malm, *Fossil Capital: The Rise of Steam Power and the Roots of Global Warming* (London: Verso, 2016).
6   Richard York and Brett Clark, "Nothing New under the Sun? The Old False Promise of New Technology," *Review: A Journal of the Fernand Braudel Center* 33, nos. 2–3 (2010): 203–224.
7   York and Clark, 203–224.
8   Brett Clark and Richard York, "Rifts and Shifts: Getting to the Root of Environmental Crises," *Monthly Review* 60, no. 6 (2008): 13–24; Brett Clark and Richard York, "Technofix: Ecological Rifts and Capital Shifts," in *Ecology and Power: Struggles over Land and Material Resources in the past, Present, and Future*, ed. A. Hornborg, B. Clark, and K. Hermele (London: Routledge, 2012), 23–36.
9   John Bellamy Foster, Brett Clark, and Richard York, *The Ecological Rift: Capitalism's War on the Earth* (New York: Monthly Review Press, 2010).
10 Brett Clark and John Bellamy Foster, "Ecological Imperialism and the Global Metabolic Rift: Unequal Exchange and the Guano/Nitrates Trade," *International Journal of Comparative Sociology* 50 (2009): 311–334.
11 Clark and Foster.
12 Philip Mancus, "Nitrogen Fertilizer Dependency and Its Contradictions: A Theoretical Exploration of Socio-ecological Metabolism," *Rural Sociology* 72 (2009): 269–288.
13 Hannah Holleman, "De-naturalizing Ecological Disaster: Colonialism, Racism and the Global Dust Bowl of the 1930s," *Journal of Peasant Studies* 44 (2017): 234–260.

14   Brett Clark and Richard York, "Carbon Metabolism: Global Capitalism, Climate Change, and the Biospheric Rift," *Theory and Society* 34, no. 4 (2005): 391–428; Malm, *Fossil Capital*; Malm, *Progress of This Storm*.

15   Keith, *Case for Climate Engineering*.

16   For my presentation of the Atlantropa Project, I relied on Alexander Gall, "Atlantropa: A Technological Vision of a United Europe," in *Networking Europe: Transnational Infrastructures and the Shaping of Europe, 1850–2000*, ed. E. van der Vleuten and A. Kaijser (Sagamore Beach, Mass.: Science History Publications, 2006), 99–128; and Richard Brook Cathcart, "Mitigative Anthrogeomorphology: A Revived 'Plan' for the Mediterranean Sea Basin and the Sahara," *Terra Nova* 7 (1995): 636–640.

17   Gall, "Atlantropa."

18   Gall, "Atlantropa." Subsequent to Sörgel's proposal, other macroengineers proposed diverting freshwater from the Amazon River to the Sahara via a floating transatlantic pipeline, building massive domes on the Sahara to help green the desert, and other breathtakingly Promethean projects. Cathcart, "Mitigative Anthrogeomorphology."

19   Gall, "Atlantropa."

20   Gall, 116.

21   Gall.

22   Gall, 100.

23   Freeman Dyson, "The Question of Global Warming," *New York Review of Books* 55, no. 10 (2008): 43–45.

24   Julia Rosen, "The Carbon Harvest," *Science* 359, no. 6377 (2018): 733–737.

25   Rosen, 733–737; John Bellamy Foster, "Making War on the Planet: Geoengineering and Capitalism's Creative Destruction of the Earth," *Science for the People*, Geoengineering Special Issue, Summer 2018, http://magazine.scienceforthepeople.org/making-war-on-the-planet/.

26   Keith, *Case for Climate Engineering*.

27   Keith, *Case for Climate Engineering*.

28   Elizabeth Kolbert, *The Sixth Extinction: An Unnatural History* (New York: Henry Holt, 2014).

13

Gramsci in the Stratosphere

Solar Geoengineering and Capitalist Hegemony

KEVIN SURPRISE

> "The crisis consists precisely in the fact that the old is dying and the new cannot be born; in this interregnum a great variety of morbid symptoms appear."
> —Antonio Gramsci, *Selections from the Prison Notebooks of Antonio Gramsci*

Driven by the layered realities of climate urgency, political inaction, and the potential for climate impacts to harm the most vulnerable, cautious considerations of solar geoengineering via stratospheric aerosol injection (SAI) are beginning to gain traction at the highest levels of climate policymaking. Yet critics contend that potentially spraying megatons of sulfur dioxide into the lower stratosphere to reflect incoming solar radiation and thus quickly cool the planet represents a particularly "morbid symptom" of capitalism's ecological crisis whereby expert elites would rather alter planetary albedo than fundamentally alter the global capitalist economy. Such critiques of SAI range from those excoriating its Promethean hubris,[1] technocratic or authoritarian potential,[2] capacity to expand fossil capital in the midst of climate crisis,[3] offer for a temporary work-around for the climate-capitalism contradiction more broadly,[4] and expansion of capitalist modernity's domination of nature.[5]

Despite broadly shared perspectives rooted in critical political economy, there are key differences between these approaches, with potentially significant politico-analytical consequences for critiques of—and challenges to—solar geoengineering. To sharpen these critiques of SAI (and geoengineering more broadly), critics need to clarify positions on three central questions: How do we understand the crisis to which SAI is emerging as a response? What is SAI

189

intended to accomplish in the context of this crisis? Given this context, which actors or institutions are most likely to develop and deploy SAI? Answers to these questions can portend vastly different approaches to SAI. For example, compare two approaches that begin from similar theoretical positions but ultimately produce quite different critiques: some of my own recent work and Andreas Malm's discussion of SAI in *Fossil Capital*.[6] This discussion is necessarily brief, and I am not attempting to set Malm's up as a straw argument (I find his analysis of the origins and expansion of fossil capital innovative and convincing). I merely intend to illustrate how different approaches to these questions lead to differing political conclusions.

Malm's analysis posits that industrial capitalism is fundamentally predicated upon and inextricable from ever-greater fossil energy use, and thus SAI is emerging as a "centrally placed emergency brake" to pull an otherwise unstoppable system back from the brink of climate catastrophe (and enable continued accumulation through the [infra]structures of "fossil capital"). This scenario envisions a toxic relationship whereby increased burning of fossil fuels necessitates ever-greater amounts of SAI in perpetuity, increasing the potential for so-called termination shock (where rising CO_2 emissions under an SAI program bottle up the greenhouse effect and, if SAI were ever halted, unleash rapid warming). Here the crisis concerns the inability—or refusal—of fossil-dominated capital to manage climate change, potentially necessitating dangerous, emergency forms of ever-escalating solar geoengineering in perpetuity to maintain and expand fossil capitalism.

This scenario is certainly possible. Nevertheless, my approach operates under the assumption that capital—particularly capitalist states—can potentially effectuate and function within a broad transition to decarbonization.[7] Yet such a shift toward so-called green capitalism is nascent, ad hoc, and running up against myriad obstacles, three of which I consider central: the interwoven emergence of increasing rates of climatic change, climate-driven costs and losses that undermine accumulation, and burdensome costs required for decarbonization. In this context, SAI to slow the *rate* of change (not offset all warming) could function as a spatiotemporal fix, temporarily elongating the timescales within which "green capitalism" becomes possible, thus reducing rate-dependent costs and losses and the immediate financial burdens of green sociotechnical transitions. Here the crisis turns not merely on ecological catastrophe but on climate-driven costs to capital that undermine the capacity to generate surplus value (the "second contradiction" thesis[8]), and thus "business as usual" concerns the capacity to effectuate decarbonization while maintaining the hegemonic structures of liberal capitalism. Hence both analyses are rooted in Marxist theory but present different understandings of the relationships between capitalism, environment, energy, and—most explicitly—the capacity for capitalist states (broadly conceived) to manage crises in the interest of liberal-capitalist hegemony.

Whether SAI is understood as an emergency tool that can enable the expansion of fossil capital—masking its climate effects in increasingly risky escalations—or a more moderate attempt to maintain the hegemony of liberal capitalism by "buying time" for the emergence of so-called green capitalism, it has significant politico-analytical implications. Myriad perspectives will be brought to bear on these questions as geoengineering debates evolve. In what follows, I offer one approach rooted in Antonio Gramsci's conceptions of "organic crisis" and "passive revolution." Gramsci defined organic crises as deep, structural crises that pose fundamental challenges to the continuity of a hegemonic order. Such crises can be managed via passive revolutions—political maneuvers undertaken to maintain hegemony in the midst of crisis—which Gramsci analyzed through three "moments or levels" (material levels of development, political forces and relations of hegemony, and politico-military struggles). I begin by elaborating Gramsci's concepts and proceed by placing SAI in the context of Gramsci's three "levels," examining the underlying socioecological contradictions spurring its emergence, the forms of governance envisioned by leading centers of research, and the military interests and geopolitical rationales conditioning potential deployment. I seek to situate SAI as a key tactic in a broader "green capitalist" passive revolution that aims to manage the ongoing organic crisis of liberal capitalism.

Crises, Passive Revolution, and Gramscian "Method"

Gramsci conceives of organic crises as deep, fundamental crises of the ruling order that cannot be resolved from within the existing political parameters of that order—"crises of authority . . . precisely the crisis of hegemony, or general crisis of the State."[9] Organic crises do not refer to the regularized crises encountered by capitalism as a crisis-ridden system. They are not merely conjunctures; rather, they *form the terrain* of the 'conjunctural,' and it is upon this terrain that the forces of opposition organize."[10] That is, contradictions of a ruling order come to an inflection point such that they pose an existential threat to the ruling class, creating openings for the "forces of opposition" to challenge and abolish existing forms of hegemony. Indeed, organic crises are "dangerous" in a dual register: First, the crisis erodes the legitimacy of the ruling class, rendering their rule one of domination rather than hegemony that is increasingly reliant on coercion and violence. Second, the radically contingent nature of crisis opens new spaces and opportunities for resistance to and abolition of the hegemonic order. In this context, Gramsci sought to understand how ruling classes maintain and reestablish hegemony in the midst of crisis to better inform revolutionary strategy. The analysis is thus necessarily pessimistic: "The traditional ruling class, which has numerous trained cadres, changes men and programmes and, with greater speed than is achieved by subordinate classes, reabsorbs the control that was slipping from its grasp. Perhaps it may make sacrifices, and expose itself to an uncertain future by demagogic promises; but it retains power, reinforces it for the time being."[11] Liberal-capitalist elites have the capacity to orchestrate shifts in contours of power that—while potentially unstable or temporary—ultimately reinforce the ruling order.

The "reabsorption" of control is an integral aspect of the passive revolution, or the strategies and mechanisms through which the ruling class enacts "revolutionary reforms" to maintain hegemony. Gramscian philosopher Peter

Thomas argues that the guiding thread of Gramsci's carceral research can be characterized as "the search for an adequate theory of proletarian hegemony in the epoch of the 'organic crisis' or the 'passive revolution' of the bourgeois 'integral state.'"[12] In other words, Gramsci's central project turned on strategizing for radical change in periods of organic crisis whereby ruling classes seek to (re)secure hegemony via passive revolutions in political and civil society. There is considerable debate over how Gramsci mobilized the passive revolution concept and its ability to "travel" beyond the contexts for which he deployed it (e.g., bourgeois revolution in France from 1789 to 1870 or Italian Risorgimento in the nineteenth century).[13] Here I find Domenico Losurdo's description of Gramsci's conception compelling: passive revolution "denote[s] the persistent capacity of initiative of the bourgeoisie which succeeds, even in the historical phase in which it has ceased to be a properly revolutionary class, to produce socio-political transformations, sometimes of significance, conserving securely in its own hands power, initiative and hegemony, and leaving the working class in their condition of subalternity."[14] The capacity for the ruling class to produce significant political transformations from within its own "power, initiative and hegemony" to maintain its rule in the midst of structural crises defines the "passive revolution."

This is a broad definition. Gramsci framed it specifically as a methodological approach:[15] an "interpretive criterion" for analyzing "molecular changes which in fact progressively modify the preexisting composition of forces, and hence become the matrix of new changes," also understood as a process of "revolution/restoration."[16] Gramsci couches this methodological approach to passive revolution in what he terms the "two fundamental principles of political science" derived from Marx's *Preface to the Critique of Political Economy* (1959), which argues that a ruling order will remain dominant until the material forces undergirding it are exhausted.[17] Gramsci draws from these passages throughout the *Prison Notebooks* yet demands that they be "developed critically in all their implications, and *purged of every residue of mechanism and fatalism. They must be referred back to the description of the three fundamental moments into which a 'situation' or an equilibrium of forces can be distinguished...*"[18] Gramsci's conception of a "situation" or equilibrium of forces is composed of three interrelated "moments or levels." First, there is the "relation of social forces" (for which we can say socioecological forces), which corresponds to "material development" measured by the "physical sciences."[19] Second, we have the "relation of political forces" (for which we can say hegemony), which is fully developed when the ruling class is able to establish their narrow interests as universal: "The development and expansion of the particular [hegemonic] group are conceived of, and presented, as being the motor force of a universal expansion, of a development of all the 'national' energies."[20] Third, there is the relation of "politico-military forces" that "from time to time is directly decisive."[21] Analyzing the layered interplay among material-economic conditions, state power, and geopolitical contestation in an era of organic crisis can inform understandings of likely responses by hegemonic actors and thus strategies for resistance.

I utilize Gramsci's "levels" as a prism through which to situate SAI as a singularly unique tactic in the broader passive revolution of "green capitalism." In its capacity to both precipitate systemic financial instability and embolden radical movements for climate justice that place the hegemony of liberal capitalism squarely in their sites, climate change is the leading front in an emergent organic crisis.[22] The primary crisis management strategy thus far advanced by capital and technomanagerial elites is so-called green capitalism: ad hoc attempts to transition to sustainability via renewable energy, carbon markets, privatized conservation, natural capital, green consumerism, carbon capture and storage (CCS), bioenergy, and so on without fundamentally altering capitalist forms of class domination, exploitation, and accumulation.[23]

Thomas Wanner explicitly couches the "green economy" as a passive revolution intended to both open new realms of commodification and neutralize radical environmentalism.[24] Similarly, Ulrich Brand and Markus Wissen argue that the "green capitalism" strategy has the potential to become a "hegemonic capitalist project."[25] This is no mere greenwashing or niche market strategy. James McCarthy argues that a broad transition toward renewable energy—while facing myriad obstacles—has the potential to open vast new realms of investment and stave off both financial crises (overaccumulation) and ecological crises (underproduction).[26] Following Gramsci's methodology, I situate green capitalism as a response to the "material relations" of contemporary socioecological crises defined by climate change. I then turn to SAI as an elite expert governance intervention—Gramsci's level of "political relations"—that offers a unique fix for this crisis. Finally, I briefly examine the geopolitical implications of this intervention and explore the role the U.S. military might play in the development of this technology.

Crisis, Green Capitalism, Temporal Disjuncture

The contours of the capitalist climate crisis are not agreed upon even by its critics. For example, perspectives from metabolic rift theory, world-ecology, and the second contradiction thesis aver that capital accumulation is the motive force of ecological destruction but differ significantly on what socioecological crises mean for the future(s) of capitalism. Proponents of the "metabolic rift" argue that capitalism's separation of producers from the means of (agrarian) production via mechanized agriculture, driven by logics of competitive growth and concentration of large-scale industry, produces a socioecological rift: depleting and degrading soil in the countryside while transporting nutrients to urban centers where they are not returned to the soil but accumulate as waste/pollution.[27] Brett Clark and Richard York extend this to the biosphere: the expansion of capitalist industrialization engendered both widespread burning of fossil fuels and extraction from ever-greater resource frontiers, producing a rift in the carbon cycle via overaccumulation of carbon dioxide in the atmosphere and degradation of natural carbon sinks.[28] This transgression of key planetary boundaries can never be repaired within the capitalist mode of production, as "capital is dictated by the competition for the accumulation of wealth . . . in this capitalism successfully conquers the earth (including the atmosphere), taking its destructive field of operation to the planetary level."[29] Thus metabolic rift theory holds that capital's solipsistic drive for accumulation will inevitably engender ecological catastrophe.

Yet capital cannot merely continue to accumulate in the midst of socioecological crisis. Jason Moore's world-ecology perspective argues that such ecological degradation threatens the production of value central to accumulation. Moore argues that the capitalist "law of value" is defined by a dialectic of appropriation and exploitation, the latter dependent upon the capacity to appropriate "Cheap Nature." Cheap labor power, food, energy, and raw materials generate ecological surpluses that subtend innovations in commodity production. Every great era of accumulation has been predicated upon new frontiers that provide an ecological surplus, enabling the appropriation of nature's unpaid work to proceed faster than the rate of exploitation. Yet the temporalities of the capitalist value-form exhaust zones of appropriation, necessitating the capitalization of once freely appropriated natures, raising costs and precipitating crises. Moore suggests that such exhaustion-capitalization is reaching a precipice in the contemporary crises of neoliberalism, signaling the end of the frontier. The double burden of resource exhaustion and toxification is forcing capital to spend more to extract less—a process Moore terms "negative-value," defined as "accumulation of the limits to capital in the web of life that are direct barriers to the restoration of the Four Cheaps."[30] In this conjuncture, Moore sees the looming collapse of capitalism under the weight of its own contradictions.[31]

Moore's concept of negative-value owes much to James O'Connor's "second contradiction of capitalism" thesis, wherein overaccumulation crises (the first contradiction) compel capitals to search for cheaper raw materials and energy to drive down costs.[32] This strategy can be frustrated by the biophysical properties of nature, as "nature's productivity is self-limiting," yet for capital, these limits are merely "barriers to overcome."[33] The second contradiction thus emerges in the drive to overcome the limits of nature's productivity: enfolding nature into the logics of exchange value negates its biophysical properties and tends to underproduce (degrade, exhaust, pollute) the "conditions of production." Underproduction can generate systemic crisis, as production conditions are fundamental to the creation of surplus value: "Favorable natural conditions increase the productivity of labor, hence *reduce* (not increase) the exchange value of commodities, and in turn (everything else being the same) increase the production of surplus value and profit."[34] Thus overaccumulation and underproduction crises are dialectically linked.

The central divergence between O'Connor and Moore turns on the capacity for capitalist states to manage ecological crises. For Moore, ecological exhaustion driven by the appropriation-exploitation dialectic augurs the end of capitalism. Yet O'Connor places primacy on the political qualities of capitalist crises, allowing for the possibility that state and class forces can manage crises (if not "solve" them). O'Connor argues that the state mediates socioecological crises as a social relation defined by multiple, competing material interests, with the definition and management of "ecological crises" turning primarily on politico-ideological questions: "'Ecological crisis' is as much (or more) a political and ideological category as it is a scientific construct."[35] Whether or not capital faces "'external barriers' to accumulation, including . . . in the form of new social struggles over the definition and use of productive conditions . . . ," and whether or not these "'barriers' take the form of economic crisis; and whether or not economic crisis is resolved in favor of or against capital are sociopolitical and ideological questions first, and socioeconomic questions only secondarily."[36]

That is, while specifically capitalist ecological crises emerge via politico-economic costs incurred through biophysical exhaustion, the key struggles revolve around whether and how the state (broadly conceived) recognizes and defines the scope of the crisis and is compelled (or not) to marshal powers of regulatory, financial, or technological crisis management. Climate change is the leading front of the emergent second contradiction, and the central management strategy advanced thus far is "green capitalism." While this emergent capitalist formation is possible, the necessary transitions will take several decades as well as several decades to prove climatically effective, all the while necessitating active coordination on the part of finance capital and highly regulative states investing tens of trillions of dollars. Thus a central variable thus determining this strategy's effectuality is time: Can it prove effective faster than the climate change comes to threaten capitalist (re)production? Here I provide a brief, illustrative list of threats to financial stability posed by climate change, juxtaposed with massive costs of decarbonization and the increasing rate of climatic change (what I call the "temporal disjuncture" of climate politics).

A warming world poses significant threats to the global capitalist economy.[37] Global mean temperature estimates, however, provide only a partial glimpse into time scales necessary for adaptation, as Patrick Smith and his colleagues argue: "faster *rates* of change [result] in less time for human and natural systems to adapt."[38] Examining rates of change over forty-year periods for the last millennium, the average rate for the nine hundred years prior to the twentieth century was 0.1°C, whereas in the latter half of the twentieth century, rates of change (for the Northern Hemisphere) have reached 0.2°C. Using Intergovernmental Panel on Climate Change (IPCC) Representative Concentration Pathway (RCP) 4.5, they demonstrate that rates of change for forty-year periods from 2030 will span 0.15°C per decade if climate sensitivity is low to somewhere between 0.33°C and 0.4°C per decade if sensitivity is high.[39] Higher climate sensitivity thus threatens to wipe considerable value off global GDP and financial markets. Using the common range of 1.5°C to 4.5°C rise in preindustrial temperature by the end of the century, the Organization for Economic Cooperation and Development (OECD) estimates that climate-driven losses in GDP could range from 1.0 to 3.3 percent per annum by 2060.[40] The greatest losses occur in crop yields and labor productivity as well as urban flooding, loss of ecosystem services, and high-impact weather events.[41] Citigroup couches this in monetary terms, estimating total losses in GDP between $20 trillion[42] (1.5°C) and $72 trillion (4.5°C) by 2060.[43]

More specifically, Simon Dietz and colleagues examine the "value at risk" (VaR) to financial assets: VaR in a business as usual (BAU) climate scenario (2.5°C above preindustrial levels by 2100) is 1.8 percent of global financial assets ($2.5 trillion).[44] The most severe risk concerns higher sensitivity, emerging in the tail end of the estimates: the ninety-ninth percentile of VaR in a BAU is 16.9 percent or $24.2 trillion.[45] Yet the costs and requirements of a rapid transition away from fossil fuels are seemingly prohibitive. For example, Mark Jacobson and colleagues argue that to achieve the Paris Climate Agreement's goal of

1.5°C by 2100, 139 countries must shift to 80 percent renewable energy (wind, water, and solar) by 2030, increasing to 100 percent by 2050, costing an estimated *$124 trillion*.[46] Juxtapose this to the International Energy Agency's (IEA) 2016 World Energy Outlook, where (under BAU) by 2040, 60 percent of total investment continues toward fossil fuels, 20 percent to renewables, and the remainder to improve efficiency.[47] Hence Climate Action Tracker's most recent estimates put current climate trajectories at 3.4°C by 2100.[48]

Climate change thus portends a systemic underproduction crisis (the second contradiction) par excellence: despoliation (overuse) of the atmospheric carbon sink may destabilize crucial socioecological functions underpinning capital accumulation, raising costs and undermining the capacity to generate surplus value. While potentially profitable in the long term, repairing the conditions of production requires onerous upfront investment costs with the capacity to drain surpluses, further threatening the (re)production of surplus value. Thus in its capacity to quickly and effectively slow the *rate* of change, solar geoengineering via SAI potentially offers a singularly unique "fix" for this emergent contradiction. Here I turn to the leading center of research developing the "limiting rates of change" approach to SAI, Harvard University.

Governing the Rate of Change

The intellectuals are the dominant group's "deputies" exercising the subaltern functions of social hegemony and political government.
—Antonio Gramsci, *Selections from the Prison Notebooks of Antonio Gramsci*

Much has been said about the potential governance of SAI.[49] My concern in this section is not any specific proposal for governance but select models and scenarios for deployment, the researchers creating the models, the people funding these researchers, and the ways in which these SAI models might work to maintain capitalist hegemony as a planetary-scale spatiotemporal fix. The leading center of such research and model making is the Harvard Solar Geoengineering Research Program (HSGRP), composed of an interdisciplinary team of researchers studying the science, governance, and experimentation of SAI—with the first field experiments (ScoPEx) likely going live in 2020.[50] Housed within the Harvard University Center for the Environment, with connections to the Kennedy School of Government and Belfer Center for Science and International Affairs, HSGRP is directed by physicist David Keith and backed by approximately $16 million in philanthropic funding from a range of foundations and private billionaires with deep ties to Silicon Valley and Wall Street.

This is an atypical constellation of funders that includes, for example, the William and Flora Hewlett Foundation; the J. Baker Foundation, the philanthropic arm of one-half of Baker Brother Advisors, an investment firm focused on biotech and pharmaceuticals; Teza Technologies, founded by Misha Malyshev, PhD in astrophysics from Princeton turned investor with Mickinsey and Company and Citadel Investment Group; the Open Philanthropy Project, founded by Dustin Moskovitz (a cofounder of Facebook) and Cari Tuna (with backing from Holden Karnofsky, formerly of Bridgewater Capital), which has donated a sizeable $2.5 million; the Pritzker Innovation Fund, run by Rachel Pritzker (heir to the Pritzker hotel fortune), who is also a major backer of the Breakthrough Institute; Alan Eustace, former senior vice president of knowledge at Google; G. Leonard Baker Jr., partner at Sutter Hill Ventures, a Silicon Valley venture capital firm; Bill Trenchard, partner at First Round Capital; Ross Garon, managing director of Point72 Asset Management; and Bill Gates, who matches 40 percent of each gift.[51]

The cross section of technology firms and hedge fund managers are drawn to the technofix aspect of SAI. As senior HSGRP researcher explains, "[SAI] is something that Silicon Valley entrepreneurs, in fact, like . . . that folks in Silicon Valley can understand to a point. As in there's a technological intervention that's not just tech, right? It requires, sort of, all three: Silicon Valley, Washington, and science. On the other hand, there's this seemingly intractable problem out there and now there is some sort of not quite standard solution being talked about."[52]

The "nonstandard" solution proffered by Harvard researchers concerns the utilization of SAI technology to slow or limit the rate of climatic change. In this approach, SAI cannot act as a substitute for emissions cuts; rather, if deployed, it can slow the rate of climatic change. For example, Keith and Douglas MacMartin lay out a scenario in which SAI deployment is moderate (does not offset all warming), temporary (end point is zero), and responsive (adjusted per information): starting in 2020 with 0.035 MtS/yr (megatons of sulfur per year) delivered to the lower stratosphere by two modified commercial jets, ramped up to 0.35 MtS/yr by 2030, and thereafter requiring a total of thirty jets by 2040, costing approximately $2.2 billion over the initial thirty-year period. This strategy can significantly delay the point at which the 2°C threshold is reached—from 2055 to 2068—and adds thirty-two years to the time frame for reaching an increase of 2.5°C.[53] Keith and Peter Irvine construct a scenario in which reducing total mean warming by half in the twenty-first century could reduce sea level rise by 20 percent and increase yields in over 70 percent of agricultural areas.[54] The rationale for SAI is thus fairly straightforward: it can quickly—and cheaply—slow the rate of change, producing net-positive global impacts that could save lives (and capital): "Rates of change are . . . an important factor in human impacts from climate change. At a minimum, introducing a delay in reaching any particular temperature-dependent climate impact would increase the amount of time to both learn and adapt; by reducing the needed rate for adaptation, *it could reduce economic costs of adaptation* . . ."[55]

Designed as such, SAI could act as a planetary-scale spatiotemporal fix for the second contradiction of capitalism.[56] As conceived by David Harvey, spatiotemporal fixes concern the ways in which capital attempts to solve crises of overaccumulation by absorbing capital surpluses via investment in long-term projects that "defer the re-entry of capital values into circulation into the future."[57] This primarily occurs through spatial displacement by opening up new markets, establishing new production capacities, locating new resource frontiers, and so on. These temporal and spatial strategies converge in various spatiotemporal fixes, as investments in the secondary sector (large-scale fixed capital assets such as industrial plants, highways, and communication networks) and tertiary sector (social infrastructures such as education, health care, and scientific research) become "fixed" on the land for long periods of time, absorbing surpluses and ensuring accumulation into the future. Hence spatiotemporal fixes offer particular solutions to capitalist crisis "through temporal deferral and geographical expansion."[58] In other words, they do not solve the internal contradictions of capitalism but delay them in time through the production of space.

SAI holds the potential to defer the onset of climate-driven *underproduction* crises through the production of atmospheric space. Spraying, for example, 0.35 MtS dioxide into the lower stratosphere yearly would temporarily alter planetary albedo, in effect expanding the atmosphere's capacity to absorb carbon dioxide without attendant warming. This intentional modification of albedo—and artificial expansion of the atmospheric waste sink—is intended to decrease the rate of climatic change. As the Harvard models suggest, SAI can potentially cut the rate of warming in half by 2100 or add decades to the timescales in which catastrophic temperatures are reached (e.g., thirty-two years to BAU 2.5°C). This temporal shift can elongate the timescales within which green capitalism proves effective, enabling more gradual energy transitions and thus drastically reducing the burdensome upfront costs of decarbonization. A similar logic adheres in relation to costs and losses: for example, in the financial sector, potential losses are rate/sensitivity dependent (e.g., $24.2 trillion if climate sensitivity is high), and overall GDP is placed at greater risk with higher temperatures (3.3 percent per year under 4.5°C). Intentionally managing the rate of change can thus potentially decrease systemic risk and volatile unpredictability. In other words, if climate change is the most expressed manifestation of the second contradiction and green capitalism the central crisis management strategy, then SAI—in producing (artificial) space through an expansion of the atmospheric waste sink via albedo modification—can defer the onset of this contradiction. This would allow for not only timely effectuation of green capitalism and the deferral of climate crisis but the expansion of capital accumulation in an otherwise finite system.

Securing a Changing Climate

Even such a "moderate" deployment of SAI would mandate considerable forms of governance, requiring annual dispersal, consistent monitoring, adjustment, compensation mechanisms, and so on. Questions of SAI governance occupy much of the literature. Here I focus on theories of international relations that inform the bulk of this literature and explore the prospects for militarization of SAI, which has been significantly underexplored in governance debates. Militarization is overlooked because the geoengineering governance "imaginary largely follows the so-called neo-neo debate in . . . mainstream US-centered international relations."[59] For neorealists, leading powers could create an international regime of rules for SAI; for neoliberals, rational state behavior could lead to cooperation and institutionalization of governance.[60] Yet both sides of this "debate" adhere to the basic structural tenets of mainstream realist international relations (IR): Territorially sovereign states as rational actors are the relevant units of analysis. Rational states compete for relative power in an anarchical international system where the potential for, or obstacles to, cooperation is the key concern.

This is evident in one of the most common tropes surrounding the international politics of SAI: its so-called free driver problem structure. The free driver metaphor is in contradistinction to the notion of free riding, or collective action problems that characterize contemporary treaty-driven climate politics.[61] SAI exhibits "free rider" characteristics because it is likely both inexpensive and high leverage such that a number of actors can potentially deploy the technology unilaterally, and the actor that desires the most cooling sets the parameters of engagement. Yet much recent literature posits that the technological and political requirements for legitimate deployment lend SAI a "logic of multilateralism" wherein cooperation and coordination among actors are essential.[62] Here the primary actors are states, and they can best serve their rational self-interest by engaging in positive-sum cooperative governance (and this form of governance may be pluralistic or dominated by a handful of powerful states).

The vast majority of the solar geoengineering governance follows the strict confines of the mainstream "neoneo" framework and thus misses crucial geopolitical questions. Realism (and its *neo-* variants) begins with a "metaphysical commitment" to core axiomatic concepts—territorial sovereignty, states as rational actors, and international anarchy[63]—that obscure the *social* roots of global power. This, for Robert Cox, renders traditional IR a "problem-solving theory"; that is, "it takes the world as it finds it, with the prevailing social and power relationships and the institutions into which they are organized as the given framework for action."[64] Realism served a "problem solving" function for American IR during the Cold War, "which imposed the category of bipolarity upon international relations, and an overriding concern for the defense of American power as a bulwark of the maintenance of order."[65] In this context, three trans- (or a-) historical realities were postulated as universal: human nature is defined by an unceasing Hobbesian desire for power, nation-states concretized this desire in a uniform conception of "national interest," and these power relations are adjudicated through the international balance of power. These "underlying substances" of (neo)realism fed into game theory, wherein states are believed to operate with a "common rationality" within the international system.[66] Against this ahistorical, flat conception of power, Cox poses a historical materialist approach to IR inspired by Marx and Gramsci in which international conflict is a product of the social relations of capitalism (what Cox calls the "state/society complex"). This approach focuses not on political power in general but on the specific geopolitics of capitalist imperialism, which adds a "vertical dimension of power to the horizontal dimension of rivalry . . . which draws the almost exclusive attention of neorealism."[67]

Given the theoretical limits and presuppositions of much of the mainstream SAI governance literature, the question of the potential for militarization of SAI is rarely broached, a case in point being a recent overview of solar geoengineering governance by Jesse Reynolds, hailed by Keith as "the best book-length examination of the governance of solar geoengineering to date."[68] In his chapter on a "path forward" for solar geoengineering governance, Reynolds argues that military involvement in solar geoengineering development should be restricted. Yet he notes that militaries also have necessary equipment and expertise such that rejection of their involvement could be harmful, thus boundaries should be established, but there is currently no mechanism capable of enforcing said boundaries.[69] This leaves the door wide open for military involvement. Indeed, Reynolds's argument misses two key aspects of militarization and SAI: First, militaries—particularly the U.S. military and the wider national security establishment—are *already* involved in SAI development. Second, while Reynolds notes the lack of tactical utility, the primary military rationale for potentially deploying SAI would be *strategic*.

First, there has been a slow but steady stream of recent military and intelligence interest in SAI. I can only offer an illustrative list here. For example, defense-oriented physicists Edward Teller and Lowell Wood began analyzing SAI deployment while at the Lawrence Livermore National Laboratory in the 1990s.[70] The Department of Defense included solar geoengineering options in their 2003 report on climate security.[71] The Council on Foreign Relations—a bastion of the U.S. foreign policy elite—held workshops on unilateral deployment in 2008.[72] The Defense Advanced Research Projects Agency (DARPA) held a meeting on solar geoengineering in 2009 and later began funding research at the National Center for Atmospheric Research and the Pacific Northwest National Laboratory (PNNL).[73] Defense-related research institutes—the Aerospace Research Corporation and the RAND Corporation—wrote reports on SAI.[74] The Center for Naval Analysis included SAI scenarios in recent "climate war games,"[75] the Central Intelligence Agency funded the 2015 National Academy of Sciences (NAS) reports on geoengineering, and former CIA Director Brennan has openly endorsed SAI research for security in a speech to the Council on Foreign Relations in 2016.[76]

This list could go on for some time, but I will end with a recent scenario produced by HSGRP that examines the costs and requirements for the first fifteen years of an SAI deployment program. In this scenario, deployment begins in 2033 with the dispersal of ~0.1 MtS, increasing at a rate of ~0.1 MtS/yr^{-1} linearly thereafter, deployed from latitudes 15° N/S and 30° N/S at an altitude of ~20 km. They run through options for delivering a large sulfur payload from multiple injection sites at high altitudes, noting that their "main research" on this question "involved engaging directly with commercial aerospace vendors to elicit what current near-term technology platforms can achieve at what cost. We have met or corresponded directly with: Airbus, Boeing, Bombardier, Gulfstream, GE Engines, Rolls Royce Engines, Atlas Air, Near Space Corporation, Scaled Composites, The Spaceship Company, Virgin Orbit, and NASA."[77] They reach the conclusion that requisite (and affordable) SAI deployment can only be achieved through the modification of existing military aircraft, which they term SAI launcher (SAIL). Maximum deployment by 2047 (in this scenario) will require ninety-five SAIL aircraft flying 41 flights per day (60,109/yr) from four "bases" (at latitudes 15° N/S and 30° N/S) to deliver 1.5 million tons of sulfur dioxide to the lower stratosphere. With modified military aircraft flown from "bases" and the involvement of major aerospace and defense contractors, this scenario is *explicitly* militarized[78] for an HSGRP scenario whereby "counter-geoengineering" is explored as a deterrent to unilateral deployment.

The increasing militarization of climate change[79]—and now of SAI—renders the capacity for cooperative solar geoengineering governance without the express involvement and consent of the United States deeply suspect. As Paul Nightingale and Rose Cairns argue, it seems highly unlikely that the U.S. military (let alone the U.S. Congress) would allow the capacity to intervene in the climate system fall to another actor.[80] If a changing climate has been deemed a security threat, it holds that any *deliberate* attempt to change the climate would, ipso facto, be considered a security threat. Further, beyond merely managing security threats, Melinda Cooper argues that SAI deployment might advance broader, *strategic* goals of the U.S. military. Cooper argues that the militarization of climate change has little to do with the climate directly; rather, climate change raises concerns for U.S. dominance in energy markets, finance, and broader geopolitical strategy: "The failure of US intervention in the Middle East, the rising fortunes of China, and the recent financial crisis have all underscored the fragility of an imperial politics pivoting around the dollar-oil nexus. Within current US strategizing then, climate change has become a security concern not only because it poses political risks of its own . . . but also because it is inseparable from the larger issues of energy security, energy transition, and the exceptional role of the US dollar within financial flows."[81]

In this context, solar geoengineering might become increasingly appealing to U.S. security planners. While it cannot eliminate climate change, it can potentially preempt the worst effects and enable the semicontrolled management of climatic change: "US strategy is less concerned about the effects of climate change as such (turbulence is assumed) than it is about the consequences for US imperial power. The precise when and where of turbulence is indifferent. What matters is whether the accidental event of turbulence can be harnessed to the strategic ends of the US-dollar denominated world."[82] That is, if the climate is inevitably changing and this change presents a threat to U.S. power (primarily by forcing a rapid shift away from fossil fuels, and particularly oil, to which the dollar is inextricably linked), then strategic logic suggests actively intervening in and changing the climate in such a way that is strategically beneficial to U.S. imperial power.

Picking up this thread, Joel Wainwright and Geoff Mann include U.S. military–directed geoengineering in their conception of "planetary sovereignty" whereby the political contours of climate change have the potential to force a reconfiguration of sovereign power from the nation-state to the planetary scale. Constructed to "allow capitalist elites to stabilize their position amidst planetary crisis," planetary sovereignty will be defined "by an exception proclaimed in the name of preserving life on Earth . . . a collection of powers coordinat[ing] to 'save the planet' and determine what measures are necessary and what and who must be sacrificed in the interest of life on Earth."[83] The "logics" necessary to enact sovereignty on such a scale turn on responding to the climate crisis via hegemonic military capacity, the production and protection of geoengineering technologies, and the power to name the emergency (and institute legitimate institutional and technical responses in its name).[84] Developing and inculcating these capacities through and within U.S. power can potentially bolster U.S. hegemony, as "any attempt to defeat the United States militarily would also seem to unsettle the very management of life on Earth."[85] If the climate crisis poses a threat to liberal-capitalist hegemony, then the U.S. military—long the enforcer of global capitalist hegemony—has incentive (as the climate is increasingly militarized) to govern the development-deployment of SAI. Not merely in the name of national security but to secure the dominance of U.S.-led world order.

Conclusion: Unthinking the Thinkable?

The ways in which the future is imagined can shape struggles over hegemonic relations of power. It matters both how we imagine alternatives and—as I (in however limited a fashion) attempt here—how we imagine the future from the perspective of power. To this end, I have sought to demonstrate that specificity concerning the parameters of crises underlying the development of SAI, the purpose for which it will be most likely used, and the actors most likely to deploy it are key elements in imagining the political economy of geoengineered futures. Mobilizing Gramsci's conception of "passive revolution," I argue that these futures must be "imagined" in order to understand the logics, strategies, and tactics that might enable liberal capitalism to maintain hegemony in the midst of an emergent socioecological organic crisis with climate change as its most expressed manifestation.

Utilizing Gramsci's three "levels" of the relations of forces in an organic crisis, I argue that the second contradiction—emerging via climate risk to financial stability, massive costs of decarbonization, and increasing rate of change—is the crisis for which SAI is emerging as a potentially key tactic. Various forces and interests are converging around SAI research and experimentation at Harvard, where leading models suggest that the technology can be deployed to slow the rate of change, buying time for adaptation and reducing the costs of energy transitions. This would constitute a deliberate intervention to change the climate and—as the U.S. military increasingly frames a changing climate as a security threat—would likely fall under the remit of U.S. national security. This is necessarily speculative. I am merely aiming to move the debate beyond the "emergency" framework noted in the introduction, as SAI is just as likely to emerge *not* as some last-ditch emergency effort to save fossil capital tout court but rather as a more modest tactic in the imperial instantiation of "green capitalism." Indeed, the passive revolution framework demands that attention be paid to "molecular changes which in fact progressively modify the pre-existing composition of forces, and hence become the matrix of new changes."[86] This holds for the slow, ad hoc nature of "green capitalism" as well as the cautious and even mundane processes through which SAI is being researched, modeled, experimented with, and designed.

Notes

1 Stephen M. Gardiner, "Why Geoengineering Is Not a 'Global Public Good', and Why It Is Ethically Misleading to Frame It as One," *Climatic Change* 121, no. 3 (December 2013): 513–525; Clive Hamilton, *Earthmasters: The Dawn of the Age of Climate Engineering* (New Haven, Conn.: Yale University Press, 2013).

2   Timothy W. Luke, "Geoengineering as Global Climate Change Policy," *Critical Policy Studies* 4, no. 2 (2010): 111–126; Bronislaw Szerszynski et al., "Why Solar Radiation Management Geoengineering and Democracy Won't Mix," *Environment and Planning A* 45, no. 12 (December 2013): 2809–2816; Mike Hulme, *Can Science Fix Climate Change? A Case against Climate Engineering* (Cambridge, U.K.: Polity, 2014); Duncan P. McLaren, "Whose Climate and Whose Ethics? Conceptions of Justice in Solar Geoengineering Modelling," *Energy Research & Social Science* 44 (2018): 209–221; Joel Wainwright and Geoff Mann, *Climate Leviathan: A Political Theory of Our Planetary Future* (London: Verso, 2018).

3   Andreas Malm, *Fossil Capital: The Rise of Steam Power and the Roots of Global Warming* (London: Verso, 2016); Nils Markusson et al., "The Political Economy of Technical Fixes: The (Mis)Alignment of Clean Fossil and Political Regimes," *Energy Research & Social Science* 23 (January 2017): 1–10.

4   Holly Jean Buck, "Geoengineering: Re-making Climate for Profit or Humanitarian Intervention?," *Development and Change* 43, no. 1 (2012): 253–270; J. P. Sapinski, "Managing the Carbon Rift: Social Metabolism, Geoengineering and Climate Capitalism" (paper presentation, American Sociological Association Annual Meeting, Seattle, August 20, 2016); Kevin Surprise, "Preempting the Second Contradiction: Solar Geoengineering as Spatiotemporal Fix," *Annals of the American Association of Geographers* 108, no. 5 (March 19, 2018): 1228–1244.

5   Ryan Gunderson, Brian Petersen, and Diana Stuart, "A Critical Examination of Geoengineering: Economic and Technological Rationality in Social Context," *Sustainability* 10, no. 1 (2018): 269.

6   Malm, *Fossil Capital*.

7   See James McCarthy, "A Socioecological Fix to Capitalist Crisis and Climate Change? The Possibilities and Limits of Renewable Energy," *Environment and Planning A* 47, no. 12 (2015): 2485–2502; Wainwright and Mann, *Climate Leviathan*.

8   James O'Connor, "Capitalism, Nature, Socialism: A Theoretical Introduction," *Capitalism, Nature, Socialism* 1, no. 1 (1988): 11–38.

9   Antonio Gramsci, *Selections from the Prison Notebooks of Antonio Gramsci*, ed. Quintin Hoare and Geoffrey Nowell Smith (New York: International Publishers, 1971), 210.

10 Gramsci, 178 (emphasis added).

11 Gramsci, 210.

12 Peter D. Thomas, *The Gramscian Moment: Philosophy, Hegemony and Marxism* (Leiden, Netherlands: Brill, 2009), 136.

13 Adam David Morton, "The Continuum of Passive Revolution," *Capital & Class* 34, no. 3 (2010): 315–342; Alex Callinicos, "The Limits of Passive Revolution," *Capital & Class* 34, no. 3 (2010): 491–507.

14 Quoted in Thomas, *Gramscian Moment*, 147.

15 See Robert W. Cox, "Gramsci, Hegemony and International Relations: An Essay in Method," *Millennium* 12, no. 2 (1983): 162–175.

16 Gramsci, *Prison Notebooks*, 109.

17 Karl Marx, "Preface," in *A Contribution to the Critique of Political Economy* (Moscow: Progress Publishers, 1977).

18 Gramsci, *Prison Notebooks* (emphasis added).

19 Gramsci, 180.

20 Gramsci, 182.

21 Gramsci, 183.

22 For example, see Naomi Klein, *This Changes Everything: Capitalism vs. the Climate* (New York: Simon & Schuster, 2014); Terran Giacomini and Terisa Turner, "The 2014 People's Climate March and Flood Wall Street Civil Disobedience: Making the Transition to a Post-fossil Capitalist, Commoning Civilization," *Capitalism Nature Socialism* 26, no. 2 (2015): 27–45; James K. Rowe, Jessica Dempsey, and Peter Gibbs, "The Power of Fossil

Fuel Divestment (and Its Secret)," in *A World to Win: Contemporary Social Movements and Counter-Hegemony*, ed. William K. Carroll and Kanchan Sarker (Winnipeg: ARP Books, 2016), 233–249; and Leon Sealey-Huggins, "'1.5° C to Stay Alive': Climate Change, Imperialism and Justice for the Caribbean," *Third World Quarterly* 38, no. 11 (2017): 2444–2463.

23 Murat Arsel and Bram Büscher, "NatureTM Inc.: Changes and Continuities in Neoliberal Conservation and Market-Based Environmental Policy," *Development and Change* 43, no. 1 (2012): 53–78; McCarthy, "Socioecological Fix"; Sara Holiday Nelson, "Beyond the Limits to Growth: Ecology and the Neoliberal Counterrevolution," *Antipode* 47, no. 2 (March 1, 2015): 461–480; J. P. Sapinski, "Climate Capitalism and the Global Corporate Elite Network," *Environmental Sociology* 1, no. 4 (2015): 268–279; Thomas Wanner, "The New 'Passive Revolution' of the Green Economy and Growth Discourse: Maintaining the 'Sustainable Development' of Neoliberal Capitalism," *New Political Economy* 20, no. 1 (January 2, 2015): 21–41; Ulrich Brand and Markus Wissen, "Strategies of a Green Economy, Contours of a Green Capitalism," in *The International Political Economy of Production*, ed. Kees van der Pijl (Cheltenham: Edward Elgar, 2015), 508–523; Wainwright and Mann, *Climate Leviathan*.

24 Wanner, "New 'Passive Revolution.'"

25 Brand and Wissen, "Strategies of a Green Economy."

26 McCarthy, "Socioecological Fix."

27 John Bellamy Foster, "Marx's Theory of Metabolic Rift: Classical Foundation for Environmental Sociology," *American Journal of Sociology* 105, no. 2 (1999): 366–405.

28 Brett Clark and Richard York, "Carbon Metabolism: Global Capitalism, Climate Change, and the Biospheric Rift," *Theory and Society* 34, no. 4 (2005): 391–428.

29 Clark and York, 408.

30 Jason W. Moore, *Capitalism in the Web of Life: Ecology and the Accumulation of Capital* (London: Verso, 2015), 277.

31 Moore, 280.

32 O'Connor, "Capitalism, Nature, Socialism"; James O'Connor, *Natural Causes: Essays in Ecological Marxism* (New York: Guilford, 1998).

33 O'Connor, *Natural Causes*, 181.

34 O'Connor, 146.

35 O'Connor, 137.

36 O'Connor, 165.

37 World Economic Forum, *The Global Risks Report 2019*, 14th ed. (Geneva, Switzerland: World Economic Forum, 2019).

38 Patrick Smith et al., "Feasibility of Space-Based Monitoring for Governance of Solar Radiation Management Activities," in *AIAA SPACE 2010 Conference & Exposition* (Reston, Va.: Aerospace Research Central, 2010), 333.

39 Smith et al., 335.

40 Organization for Economic Cooperation and Development (OECD), *The Economic Consequences of Climate Change* (Paris: OECD, 2015), 15.

41 OECD, 12.

42 The symbol $ refers to U.S. dollars.

43 Jason Channell et al., *Energy Darwinism II: Why a Low Carbon Future Doesn't Have to Cost the Earth* (New York: Citigroup, 2015), 5.

44 Simon Dietz et al., "'Climate Value at Risk' of Global Financial Assets," *Nature Climate Change* 6, no. 7 (2016): 676.

45 Dietz et al.

46 Mark Z. Jacobson et al., "100% Clean and Renewable Wind, Water, and Sunlight All-Sector Energy Roadmaps for 139 Countries of the World," *Joule* 1, no. 1 (2017): 114.

47 International Energy Agency (IEA), *International Energy Outlook 2016: With Projections to 2040* (Paris: IEA, 2016).

48   Carbon Tracker Initiative, *Unburnable Carbon 2013: Wasted Capital and Stranded Assets* (Carbon Tracker and Grantham Research Institute, 2013).

49   For an overview, see Jesse L. Reynolds, *The Governance of Solar Geoengineering: Managing Climate Change in the Anthropocene* (Cambridge: Cambridge University Press, 2019).

50   John A. Dykema et al., "Stratospheric Controlled Perturbation Experiment: A Small-Scale Experiment to Improve Understanding of the Risks of Solar Geoengineering," *Philosophical Transactions of the Royal Society A: Mathematical, Physical and Engineering Sciences* 372, no. 2031 (2014): 20140059.

51   Interview, senior HSGRP researcher, February 27, 2018.

52   Interview, senior HSGRP researcher, March 5, 2018.

53   David W. Keith and Douglas G. MacMartin, "A Temporary, Moderate and Responsive Scenario for Solar Geoengineering," *Nature Climate Change* 5, no. 3 (March 2015): 201–206.

54   David W. Keith and Peter J. Irvine, "Solar Geoengineering Could Substantially Reduce Climate Risks—a Research Hypothesis for the Next Decade," *Earth's Future* 4 (2016): 549–559.

55   Douglas G. MacMartin, Ken Caldeira, and David W. Keith, "Solar Geoengineering to Limit the Rate of Temperature Change," *Philosophical Transactions of the Royal Society A* 372, no. 2031 (2014): 2 (emphasis added).

56   Kevin Surprise, "Preempting the Second Contradiction: Solar Geoengineering as Spatiotemporal Fix," *Annals of the American Association of Geographers* 108, no. 5 (2018): 1228–1244.

57   David Harvey, *The New Imperialism* (New York: Oxford University Press, 2003), 109; see also David Harvey, *The Limits to Capital* (London: Verso, 2007).

58   Harvey, *New Imperialism*, 115.

59   Olaf Corry, "The International Politics of Geoengineering: The Feasibility of Plan B for Tackling Climate Change," *Security Dialogue* 48, no. 4 (2017): 303.

60   Corry.

61   For a critique of the "free rider" concept, see Wainwright and Mann, *Climate Leviathan*, chap. 5.

62   See Joshua B. Horton, "Geoengineering and the Myth of Unilateralism: Pressures and Prospects for International Cooperation," *Stanford Journal of Law, Science & Policy* 4 (2011): 56–69; Scott Barrett, "Solar Geoengineering's Brave New World: Thoughts on the Governance of an Unprecedented Technology," *Review of Environmental Economics and Policy* 8, no. 2 (2014): 249–269; and Joshua B. Horton et al., "Solar Geoengineering and Democracy," *Global Environmental Politics* 18, no. 3 (2018): 5–24.

63   Richard K. Ashley, "The Poverty of Neorealism," *International Organization* 38, no. 2 (1984): 225–286.

64   Robert W. Cox, "Social Forces, States and World Orders: Beyond International Relations Theory," *Millennium* 10, no. 2 (1981): 128.

65   Cox, 131.

66   Cox, 131–132.

67   Cox, 143.

68   Reynolds, *Governance of Solar Geoengineering*, quote from jacket cover.

69   Reynolds, 207.

70   Edward Teller, Lowell Wood, and Roderick Hyde, "Global Warming and Ice Ages: I. Prospects for Physics-Based Modulation of Global Change" (paper no. UCRL-JC-128715, Lawrence Livermore National Laboratory, 1996); James R. Fleming, "The Climate Engineers," *Wilson Quarterly* 31, no. 2 (2007): 46–60.

71   Peter Schwartz and Doug Randall, *An Abrupt Climate Change Scenario and Its Implications for United States National Security* (Pasadena, Calif.: Jet Propulsion Laboratory, 2003).

72   Katharine Ricke et al., *Unilateral Geoengineering: Briefing Notes* (Washington, D.C.: Council of Foreign Relations, 2008).

73   Ben Kravitz et al., "First Simulations of Designing Stratospheric Sulfate Aerosol Geo-engineering to Meet Multiple Simultaneous Climate Objectives," *Journal of Geophysical Research: Atmospheres* 122, no. 23 (2017): 12616–12634.

74   Smith et al., "Feasibility of Space-Based Monitoring"; Robert J. Lempert and Don Prosnitz, *Governing Geoengineering Research: A Political and Technical Vulnerability Analysis of Potential Near-Term Options* (Santa Monica, Calif.: Rand Corporation, 2011).

75   Catherine M. Schkoda, Shawna G. Cuan, and E. D. McGrady, *Proceedings and Observations from a Climate Risk Event* (Arlington, Va.: Center for Naval Analysis, 2015).

76   John Brennan, "Director Brennan Speaks at the Council on Foreign Relations," Central Intelligence Agency, June 29, 2016, accessed October 10, 2018, http://www.cia.gov/news-information/speeches-testimony/2016-speeches-testimony/director-brennan-speaks-at-the-council-on-foreign-relations.html.

77   Wake Smith and Gernot Wagner, "Stratospheric Aerosol Injection Tactics and Costs in the First 15 Years of Deployment," *Environmental Research Letters* 13, no. 12 (2018): 124001.

78   Andy Parker, Joshua B. Horton, and David W. Keith, "Stopping Solar Geoengineering through Technical Means: A Preliminary Assessment of Counter-Geoengineering," *Earth's Future* 6, no. 8 (2018): 1058–1065.

79   Emily Gilbert, "The Militarization of Climate Change," *ACME: An International E-Journal for Critical Geographies* 11, no. 1 (2012): 1–14.

80   Paul Nightingale and Rose Cairns, "The Security Implications of Geoengineering: Blame, Imposed Agreement and the Security of Critical Infrastructure" (working paper series 018, Climate Geoengineering Governance, University of Sussex, 2014); see also Gardiner, "Why Geoengineering."

81   Melinda Cooper, "Turbulent Worlds," *Theory, Culture & Society* 27, nos. 2–3 (2010): 183.

82   Cooper, 184.

83   Wainwright and Mann, *Climate Leviathan*, 15.

84   Wainwright and Mann, 151.

85   Wainwright and Mann, 152.

86   Gramsci, *Prison Notebooks*, 109.

14

Promises of Climate Engineering after Neoliberalism

NILS MARKUSSON, DAVID TYFIELD,

JENNIE C. STEPHENS, AND

MADS DAHL GJEFSEN

The political landscape surrounding issues of climate engineering is currently changing fast with the rise of new nationalist populism across the United States and Europe. What do these latest developments suggest for the future envisioning, development, and deployment of climate engineering? How could this technology develop in interactive parallel with a changing political regime? And how can systematic speculation about such futures help us understand the current political turbulence?

In characteristically incoherent fashion (though aligned with other deniers), U.S. President Donald Trump has both denied the existence of anthropogenic climate change and lauded "clean coal" technology as a means of ensuring continued coal use.[1] This position is confusing in its own right. But its implications for climate policy framing and strategy are also unclear. On the one hand, it could be seen as a good fit with the established pattern of neoliberal climate change policy. Neoliberal climate policy has been characterized by reliance on emissions markets and shored up by *promises* of climate engineering technology: technical fixes to climate change—primarily carbon capture and storage (CCS) and also, more recently, explicit climate engineering through carbon dioxide removal (CDR)—and, though less so, solar radiation management (SRM). To date, climate policy in the neoliberal regime has only developed weakly performing emissions markets that do not threaten fossil-dependent industries or the implementation of climate engineering technology.[2] Marketizing the atmosphere through emissions trading has been the quintessential neoliberal stance[3] and entailed an ambiguous position that superficially acknowledges climate change as an object worthy of a policy response, while at the same time denying its reality as a systemic crisis. In this sense, the particularities of Trump's incoherence are merely an extension and enhancement of a long-standing and deeply sedimented neoliberal tension in the climate policy arena.

But there may be a more fundamental shift on the horizon. Trump has also departed from key neoliberal policies in the form of canceled trade deals and even evoked fears that his ascendancy represents a resurgent illiberalism,[4] for example, through his challenges to the judiciary in the context of immigration policy and in his open nepotism by giving prominent positions to family members in the administration. Trump has not (yet, at least) mentioned climate engineering, but there are prominent people in his administration that are long-term supporters of such technology.[5] This potential of the Trump administration to look favorably toward climate engineering resonates with warnings of inherent illiberalism of some SRM-type technologies.[6]

Trump is now the elected leader of the hegemonic country of the current world order. But his election should also be seen, together with Brexit in the United Kingdom and a growing European racist populist nationalism, as indicative of a direct challenge to neoliberal globalization from across the Global North as well as of the (longer process of) decline of U.S. hegemony.[7] In all these respects, these seemingly epochal developments can also be seen as the products of neoliberalism itself—its bastard offspring—as, of course, is Trump himself, the asset-stripping property magnate and agnotological post-truth, social media–trolling celebrity, as archetype and epitome. Trump is thus not only a continuation of neoliberalism but also entirely dependent on widespread reaction against neoliberal globalization, and so he embodies the turbulence of political-economic epochal change, signifying the transition to a new regime. Such a shift of the tectonic plates of the political economy must surely also have profound impacts on—or rather coevolve tightly with—the future of the development of a technological intervention as politically controversial and consequential as climate engineering.[8] Under this new and dynamic regime, the promises of technical fixes to climate change in the form of climate engineering (CCS, CDR, and even now SRM) are being renegotiated.

We are, it seems, living through a time of systemic turbulence when dialectical tensions are erupting and becoming manifest at an accelerated pace. At the level of transformations in the political-economic regime dominating global society (and so, for the time being, global capitalism), we see two such overlapping and interacting dialectics playing out in the late 2010s. The first of these concerns is precisely the creative destructive interplay of a still-dominant, if nondead,[9] neoliberalism and the rise of its illiberal progeny, fueled precisely by the increasingly self-destructive dysfunction of the former. This is a development equivalent—but crucially different in substance—to the global conflict of a failing British imperial liberalism and the rise of fascism in the 1920s and '30s.[10]

In the United States, at the heart of the disintegrating neoliberal world order, neoliberal certainties, such as the infallibility of "the market" that was previously politically radical and strategically effective precisely as such, are increasingly trapped in a double bind whereby they are either without opposition from "serious" opinion and/or have lost popular credibility, in particular as proposed solutions to the key issues of the day, including economic stagnation and underemployment. The politically radical dynamism of neoliberalism, upon which that regime is founded, is thus neutralized and even rendered to seem *inadequately* radical, especially as it is challenged by bold illiberal actions (even as these are themselves sometimes reined back in). This is evident in the climate policy arena, where Obama-era commitments to climate mitigation and the U.N. Framework Convention on Climate Change (UNFCCC) Paris Agreement have been dismissed with a rollback of power sector emissions regulation (the Clean Power Plan) and withdrawal from the Paris Agreement while climate denialism has been bolstered and mainstreamed.[11] A particular form of stable (if arguably stagnant) policy based on an entrenched, if contested, mix of economic dogma (climate science and promises of technical fixes) has been juxtaposed with the instability of rash political gambits, disregard for science, and heated controversy.

Disparate articulations of the future of CCS and climate engineering are part of this dynamic. In the spring of 2017, op-eds in the *Guardian* speculated about whether the Trump administration's rollback of climate policy will make climate engineering seem more urgent, whether Trump may come out in favor of climate engineering (rather than opt for full-on denialism), and whether climate engineering researchers would accept the Trump administration's support given its links to climate denialism.[12] All this, however, points to the second dialectic: namely, regarding the damage done by all this domestic political transformation in the United States to its global domination and the possible rise in its place of a new polity at the center of twenty-first-century capitalism, with the only viable candidate being China. Again, of course, this second dynamic has parallels in the 1930s, concerning the decline of the Pax Britannica and the ascendancy of a (reluctant, at time isolationist) American hegemon. We discuss this second dynamic next in terms of the still-embryonic exploration of climate engineering in China, there under very different political and sociocultural circumstances and approaches to those pursued to date in what has been an overwhelmingly Euro-American debate.

This chapter thus aims to analyze the current moment of turbulence with regard to the ongoing coevolution of the (geo)political regime and the deployment potential of climate engineering as supposed technical fixes for global climate change. We follow the discussion of the political economy of geoengineering in the preceding two chapters by Richard York and Kevin Surprise and look at future scenarios of geoengineering in various regimes. Informed by previous analyses of the cultural political economy (CPE) of promises of technical fixes to climate change,[13] we will here discuss possible coevolutions of the political regime (neoliberal, illiberal, and other scenarios) with climate engineering technology. We will also discuss the current dialectic between the lingering neoliberal regime and the illiberal upstart and between continued neoliberal U.S. hegemony and the rise of an alternative China-centered regime and how the promise of climate engineering is implicated in these dialectics. We are here concerned with how best to understand the "contemporary hinge"[14] between the past of a neoliberal regime shored up by climate engineering *promises* and multiple potential futures of the political regime coevolving with climate engineering.

The Cultural Political Economy of Promises of Technical Fixes to the Climate Change Problem

In previous work, we developed a CPE framework to analyze promises of technical fixes to the climate change problem.[15] We argue that the dominant neoliberal regime resulted in a persistent nonimplementation of CCS, CDR, and SRM. We analyze how technology, both as rhetorical promises and as substantial development and deployment, coevolves (nondeterministically) with political regimes. Promises of new technology may justify new spatiotemporal fixes for capital to invest in and hence new industries that can underpin political regimes. In turn, political regimes may favor the kinds of technology that can support their underpinning industries. As long as a contingent and emergent cycle of positive reinforcement can be maintained, technology and political regimes can coevolve in a mutually supporting, dynamically stable pattern of interaction.

Moreover, we distinguish between ordinary and defensive spatiotemporal fixes, where the latter's main benefit is one of defending the former in the face of threats. For example, while CCS technology has some potential to generate profit and growth in its own right, it is more important economically in its potential to defend existing and future fossil fuel–based operations across a range of industries in the face of the climate mitigation imperative. It is in this sense a defensive spatiotemporal fix, and the promise of CCS technology is the promise of a technical fix to the climate change problem while leaving existing sociotechnical and political-economic energy systems substantially intact.

A large and ever-growing literature has detailed how the history of roughly the past four decades, up to and possibly including the new age of Trump, has been dominated by a neoliberal political regime. By this, we mean a political regime that is at its core an epistemic project, organized around a belief in the unlimited capacity of markets to determine the optimal allocation of all things, including societal attention to ecological and other problems and the conceptual prioritization of policies that support economic growth and overcome barriers to growth.[16] These neoliberal commitments have impacted how knowledge production is organized and how it shapes social change. Specifically, the neoliberal regime has favored short-term financial profit making over long-term infrastructure investment, privately appropriable profits over public goods, and opportunistic, venturesome exploitation of (possibly existing) assets over innovation of radical new technologies.[17] The neoliberal regime has thus engendered a reorganizing of society, technology, and industry and the creation of new winners and losers that has empowered ideological cheerleaders that then support further neoliberal policies, thus setting up positive feedback loops.

Moreover, the neoliberal regime, with its commitment to limitless markets and unlimited growth, propagates an illusion of resource inexhaustibility, as exemplified by the economics of oil.[18] In dealing with climate change as a problem and potential barrier to continued economic growth, the neoliberal regime has unsurprisingly turned to economic instruments in the first instance, with emissions trading as the emblematic policy instrument. This has then been married with the promise of CCS technology, which helped fossil fuel–invested climate deniers acknowledge climate change and envision a future of both con-tinued fossil fuel use and climate mitigation.[19] The promise of CCS, therefore, has perpetuated the neoliberal political economy regime that underlies the ever-more pressing climate change imperative.[20] Moreover, within the neolib-eral political regime that promoted CCS in the 1990s and 2000s, the CCS promise helped make economic instruments like carbon taxes or carbon trad-ing palatable to countries in the Global North whose climate ambitions clashed with dependence on fossil fuel extraction—for example, the United States and Norway.[21] Internationally, the outcome of this process is that neoliberal climate policy has by and large failed and has had limited impact on industry gener-ally and on CCS investment specifically. Emissions trading schemes have either not been introduced or been allowed to generate prices too low to reduce emis-sions meaningfully. Reiterating the promise of CCS while supporting lim-ited research has been more palatable to governments around the world than actually investing in expensive CCS demonstration plants.[22] The simultaneous ongoing failure of the neoliberal regime to implement more than a few CCS facilities and continued reliance on the promise of CCS in climate modeling shows that the CCS promise has worked to support the regime precisely and only by remaining a promise.[23] Significant public investment in innovation and infrastructure would run directly counter to the policy prescriptions of CCS's most powerful—that is, neoliberal—supporters.

It is within this neoliberal context that the practical and imaginary concep-tualizations of climate engineering strategies have emerged. The trajectory of CDR appears to be following in the footsteps of CCS. As the CCS promise turned into disappointment[24] and the climate change crisis worsened, the policy gaze has turned to CDR technology, especially bioenergy with carbon capture and storage (BECCS), as Wim Carton discusses in chapter 3 in this volume. Like CCS, CDR is now positioned as a promise of a technical fix to align cli-mate mitigation with economic growth by (like CCS before it) allowing the construction of scenarios with limited warming alongside continued fossil fuel use that is largely unchanged. CDR has thus emerged as a fresh (for now at least) neoliberal technical fix promise that has yet to be checked by experience of (a lack of) implementation and as such is valuable to the regime. Like CCS, it may be most valuable to the regime as a promise, and that does not bode well for CDR development and implementation but rather suggests a repeat of the CCS interlude.

In contrast, the promise of SRM has not aligned well with the neoliberal regime. SRM has been depicted in high-profile reports as beyond the pale and surprisingly unfit for inclusion in market-based policy solutions.[25] It has been presented as both physically dangerous, with unpredictable, potentially disas-trous side effects on the climate system, and politically problematic (indeed illiberal) in that it may require globally centralized decision-making that would be hard to check democratically. It has thus served to signpost a limit to accept-able neoliberal technical fixes and so also a threat of what transgressions neces-sarily await if we don't support and implement the acceptable fixes.[26] But when relegated to being an external disciplining force, the promise of SRM has *also* challenged the foundational neoliberal belief in the unlimited ability of markets (and market instruments) to overcome barriers of change. For the very proffering of a distinction between such acceptable and unacceptable technologies is precisely to concede the independent existence of "limits"—a profoundly destabilizing eventuality for neoliberalism as a whole. The dynamic of the neoliberal regime, which admits no obstacles or constraints and revels in its destruction of "sacred cows," unfolding with repeated promises of technical fixes for climate mitigation thus seems to have met a real—and so, *as such*, potentially fatal—stumbling block. That said, SRM could yet be embraced by neoliberalism as an extreme state intervention necessary to make market solutions to climate change work. If this happened, we might then envision the introduction, in turn, of competition and privatization to extract whatever rents are possible on the back of such a state intervention. So far, though, such state intervention has been deemed excessive, and no alliance has been created backing this alignment of state SRM with emissions markets.

In previous work, we used our analytical framework to speculate systematically about possible futures resulting from plausible coevolutions of a political regime with promises of climate engineering as a technical fix to climate change. We sought to assess what evidence there was for each. Looking back, those scenarios already need updating, which tells us something about the turbulence of the current moment (just after the election of Trump, the Brexit referendum, etc.) and the unfolding evolution of both the current political regime and the climate engineering, which may be going through a moment of radical rapid change. A key rationale for writing this chapter is the illiberal tendencies of recent times and their challenge of the incumbent neoliberal regime,[27] but we are well aware that the future is more radically open-ended than that dichotomy suggests. In the following section, we will first elaborate on the range of scenarios to show some of the variety of multiple futures that are possible for coevolving climate engineering and political regimes. In the fourth section, we will use those scenarios to shed new light on current affairs.

Future Scenarios

Here we identify three distinct scenarios: (1) a continued neoliberal regime, (2) an illiberal regime, and (3) an emergent "liberalism 2.0" centered on China. These may be succinctly distinguished in terms of which fractions of global capital are dominant in each case.[28] In the neoliberal regime, globalized finance capital continues to dominate productive capital (both the automobile, consumer electrics, and so on fraction underpinning U.S. hegemony, now also globalized, and the more recently emerging digital economy). The illiberal scenario entails a retrenchment of the domination of finance capital now firmly dependent upon illiberal state power—in a new (un)Holy Alliance—and hence divided up into national factions that incubate a global geopolitical context of tension and possibly conflict. Here then, U.S. global dominance is even more flagrantly a situation of sheer "domination," not even clothed in the velvet glove of "hegemony."[29] In the liberalism 2.0 scenario, a new productive capital fraction based on the digital economy takes center stage with China, where many of these businesses are already in evidence, as a new hegemon.

For each scenario, we will both discuss the potential evolution of the political regime and speculate about the fate of climate engineering (as promise and as implementation). We draw on our analytical framework for the coevolution of technical fix promises and political regimes and our previous work on the history of neoliberalism and climate engineering. To demonstrate how each scenario might evolve from tendencies in the current turbulent state of affairs, we will also mobilize supporting evidence about the current situation for each scenario. The time frame envisioned for the scenarios to unfold is 15–30 years.

Continued Neoliberal Regime

It is possible that the current drama around Trump and Brexit ends up being only a minor upheaval that neoliberalism takes in its stride. In the United States, Trump's private economic interests may come to override his right-wing populist political positioning.[30] Several commentators suggest that his politics, especially domestically, is in fact predominantly neoliberal, with, crucially, a friendly relationship to Wall Street and the financial sector.[31] To the extent that Trump himself actually leads an administration of untrammeled Republican dominance, were he to be removed from office, a Pence administration may also shed some of the more outrageous populism and double down on just such a renewed and newly ruthless neoliberal agenda. This would involve a frank celebration of inequality but where the very Prometheanism of such "winner-takes-all" competition turns out itself to elicit breakthroughs to big new horizons of profitable investment—profiting *from*, not just *in spite of*, the current political and environmental turbulence—that enable a new wave of economic growth that relieves and/or distracts from populist grievance somewhat. For instance, in the United Kingdom, key industrial sectors, including finance, may well find ways to profit even from Brexit and help sustain the neoliberal political regime.[32]

In this scenario, neoliberalism survives the current nationalist, protectionist challenge, resulting in renewed commitment to free trade and continued globalized financialization of the economy. Since the last couple of decades have also seen spectacular growth across most of the world, with the only clear winners being the working and middle classes of the Global North, global appetite for such a regime is also eminently plausible. Deregulation and privatization thus continue to be priorities, and market instruments continue to be seen as the solution for any issues by intranational elites across the world. The "success" of a further kick start in economic growth would also serve to vindicate, and so strengthen, core neoliberal beliefs in the limitless primacy of market-based entrepreneurialism, which would, in turn, feed into a rejuvenated positive feedback loop between fossil fuels, especially oil and gas, and (belief in) unlimited growth.[33] Meanwhile, international commitments to climate change policy would likely evolve but remain ineffective and insufficient.

In light of the history of the coevolution of neoliberalism with the promises of climate engineering, with this scenario, we would expect continued nonimplementation of climate engineering. The carbon price signal would remain weak, and current promises would remain stable for some time, with continued interest in CDR technology and CCS rumbling along weakly in the background. SRM would most likely remain out of bounds, beyond perhaps some limited research, but a resurgent concentration of power in neoliberal hands could in time reopen exploration of this possibility too, now able to sweep objections aside. As this could, in turn, further concentrate forms of economic and environmental power, particularly in terms of opportunities for private profit making from the privatization of climate engineering initiatives perhaps initially introduced at state level; this could also set up new positive feedback loops of deepening coproduction with a rampant neoliberalism 2.0. The key to such a scenario, however, would be the reemergence of new dynamics through which finance capital would once again be able to profit from—and so be progressively unified and empowered by—the growth of markets, market-based technologies, technological fixes, and the problems they themselves instigate.

How this scenario may play out in the longer term as the climate crisis—or, indeed, the instability of the global economic system through ever-deeper financialization and concentration of wealth—escalates and puts pressure on the regime is unclear. As climate impacts worsen, the regime will probably face escalating carbon prices. Yet whether these are fundamentally destabilizing or simply feed further neoliberalization would depend on just how effective (global) neoliberal forces have been in further concentrating power in their hands in the meantime. After all, neoliberalism has weathered many seemingly "fundamental" destabilizations over the last few decades, precisely insofar as it can invert such conjunctures into opportunities for profit. If successful, then, high prices could conceivably be readily accommodated by the wealthy (both individuals and corporations) while simply forcing the costs on others who cannot afford them but have inadequate resources to resist further worsening of their situation; that is, a further exacerbation of inequality that may be troubling in itself but is not per se for neoliberalism. Here, then, a neoliberal "solution" to climate change may even ultimately emerge (as discussed previously) but at the cost of ever-worsening economic and political inequality. In this scenario, we could see implementation of climate engineering technologies.

Illiberalism

Another possible scenario emerges from the possibility that the illiberal tendencies of increasing numbers of national leaders go unchecked. Their appeals to nationalism and strong leadership bring about continued high levels of right-wing populist support for narrowly focused, nationalistic policies[14] emboldening the illiberal elements of the Trump administration—or, again, the Republican government that would remain in place and accountable to its angry populist base even where Trump himself is removed from office. Here we could imagine a sustained and successful strategic effort to weaken and conquer the judicial and legislative branches of government on top of the all-out assault on administrative arms of government working in the "public interest" or for social welfare, notably environment, energy, science, and education.

In this scenario, and in contrast to the neoliberal regime, the governance epistemology is not that of a market but that of a dictatorship characterized by fear, nepotism, and corrupt networks of personal enrichment playing out a populist politics of anti"liberal" outrage. The market is tolerated and perhaps even celebrated in discourse, at least domestically, but in practice, market liberalism is only allowed when it does not challenge the leader's fancies, family, and friends. It must be noted, of course, as flagged previously in terms of the "bastard off-spring," that this is in a sense an extension of neoliberalism that has long concentrated power in individuals and corporations (and networks thereof) under its banner of freeing the supposedly impersonal and disinterested market.[35] Notwithstanding this important continuity, however, the flagrant abuse of power and whipping up of sociocultural division and animosity marks an important discontinuity and difference in this scenario to that just considered previously.

The utter dominance of finance capital over productive capital is here broken but, primarily due to the weakening of the former, weighed down and split by its own internal contradictions, not the strengthening of the latter. Instead, a third actor—namely, the nation-state—steps in to fill the power void. In particular, a novel coalition emerges between illiberal forces, taking the reins of state power, and the sympathetic elements among the existing powers of a now-threatened financial elite. This coalition drives a program that aims to preserve its privileged status amid the disintegration of the neoliberal order through systematic dissimulation, concealing its elite, self-serving agenda beneath a banner of populist grievance and the sowing of division.

Such an approach would include the promise of reviving industries and associated fractions of productive capital from an erstwhile heyday (e.g., of U.S. hegemony, but consider also Russia, Turkey, and so on). Globalization would also be rolled back and national capital factions strengthened. Industry serving the military (a key priority, connecting concentrated authoritarian power and populist jingoism) or providing means of popular surveillance provides stable jobs, possibly alongside construction and realty sectors benefitting from large infrastructure investments in which the populist leadership are themselves predominant creditors.[36] We might even see the reshoring of industries.[37] Since there is already evidence of this development, what is key is the plausibility of a dynamic for its continuation over the medium term: as deepening illiberalism in government action feeds greater turbulence—which both further challenges the capacity of finance capital to fashion the world in its preferred image while also offering multiple and growing openings for opportunistic profit making (of the kind finance has perfected under neoliberalism)—thence deepened are the social and political crises that underpin further populist anger.

All in all, it is less productive as a capitalist regime, although an initial investment boom may give it a temporary boost. In perhaps marked distinction to the continued neoliberal scenario, where the growth of renewables driven by corporations and market forces continues, here society deliberately and defiantly still runs on fossil fuels, potentially with an added renaissance for centralizing nuclear power. Trump's defense of the coal industry and coal jobs[38] leads to some resurgence in coal mining and use in the United States, while all "unconventional" sources of oil and gas are developed regardless of environmental impact or even the economics of a "business case." The powerful global influence of the Trump administration through the enduring levers of U.S. power, including dollar seigniorage and unrivaled military firepower, conditions parallel the unraveling of liberal democracies throughout the world, not least through support (whether moral, as exemplar, or more actively engaged) for insurgent populisms within these countries as well.

This illiberal scenario appears compatible with the evolution of multiple different outcomes for climate engineering. Defense of coal may evolve alongside continued successful denial of climate change and therefore result in little reason to support climate engineering. But under an illiberal regime, there is also scope for autocratic use of climate engineering, especially SRM. The centralizing quality of some SRM technologies has reasonably made many observers wary of their democratic consequences,[39] and there have been warnings that climate engineering with global consequences could be undertaken not through international cooperation but by single actors. Victor warned that a wealthy "greenfinger" person could deploy some climate engineering technologies. Clearly, an autocratic leader of a rich nation could unilaterally deploy SRM technology.[40]

An unaccountable, unreliable "strong leader" could deploy SRM on a whim, but deployment could also be quite strategic. SRM could be justified by a leader or a group of leaders as a means to benevolently reduce the suffering of some of the world's most vulnerable people. However, in practice, controlling the distribution of climate impacts is surely going to be extremely challenging, perhaps impossible, and would come with a serious risk of leading to international conflict. The military potential of SRM technology[41] might also be attractive to a unilateralist nationalist leader backed by a strong military industry.[42] Climate emergency might be invoked as an excuse for developing weaponized climate engineering technology and could have the added benefit of projecting an image of strong leadership befitting a populist leader, feeding populist opinion, fear, and support on which such a regime would depend. Climate engineering technology may also be used for domestic weather-controlling reasons, but if so, it is likely to spill over into regional conflict and so again have military implications. The decision to take unilateral national climate engineering action here thus correlates with a growing geopolitical climate of great power tension and bluster, as only such military powers would (believe they) have the capacity to ignore global pressure to consider implications on other countries. Yet conversely, amid growing lack of global cooperation (including on climate change) and heightening geopolitical antagonisms (exacerbated by worsening differential climate change impacts), unilateral (and possibly weaponized) climate engineering becomes more likely.

Liberalism 2.0

The election of Trump may also signify the descent of the United States as hegemon. Trump's promises to make the United States as strong as it once was, as in his campaign slogan to "Make America Great Again," can be interpreted as widespread popular recognition, not least in the United States itself, that its global reign is weakening. The phenomenal growth and current size of the Chinese economy, including in emerging digital and knowledge economy sectors, suggest that Chinese ascendancy and the Chinese capitalist regime are critical determinants in the future world order.[43] Trump's early foreign policy has been erratic, but he repeatedly denounced China in his presidential campaign regarding its negative economic impact on the United States, and clashes with China have already emerged, notwithstanding attempts to court the Chinese president Xi Jinping. This may reflect friction in the process of global realignment.

Tyfield[44] has discussed the possibility of a qualitatively novel future liberal regime centered on China that marks a different combination of continuities and discontinuities from the neoliberal regime to those of the other two scenarios mentioned previously. This involves the turbulent emergence of a revitalized regime of the classical laissez-faire liberalism but refitted for the age of wicked problems and complex systems (and sciences thereof)—including, of course, climate change itself—as against the overweening epistemic confidence and belief in "progress" of its first incarnation in early nineteenth-century Britain. Like classical liberalism, then, this "liberalism 2.0" or "complexity liberalism" is characterized by the primacy of individualized negative liberty and a continuing privileging of markets, marking a distinct continuity with the waning neoliberal dominance of the present.

Like the power regime of classical liberalism, it is also characterized by and fueled by a dynamic of essential contestation. Here embryonic political-economic and sociotechnical disruptions (in the nineteenth century the industrial revolution, today digitization) unsettle previously stabilized political allegiances, producing a political landscape of fragmentation and new polarization. This feeds a relatively lawless growth of the new capitalist economy while the political turbulence and newly vocal "extremes" forge a qualitatively new political spectrum. This, in turn, constructs a new "reasonable" political middle of "liberal" opinion based on the increasingly empowered winners in this new world of the emergent bourgeoisie of this new capitalist economy. In short, then, under continuing and relatively unchallenged conditions of the capitalist political economy, deepening political turbulence serves only to constitute and empower a new and specifically liberal capitalist historic bloc, even as the reins of government may veer wildly to the political extremes along the way.

But there are also important differences to both of these earlier regimes. As against classical liberalism, it is no longer the sovereign individual that stands at the core of this social ontology but the intersubjectively dependent and suggestible networked individual, situated within and constitutive of contemporary complex sociotechnical systems increasingly mediated by digital technologies. An understanding of these complex systems, and their potentially turbulent and/or sudden shifts and dynamic disequilibria, thus emerges as the epistemic (and tacitly normative) basis for the argument of the prima facie superiority of unleashing market forces.

But conversely, rebasing the argument for the market in a newfound body of thought also sets this regime directly against neoliberalism. For this move deposes markets per se from their foundational (and fundamentalist) centrality and reestablishes politically and scientifically compelling arguments regarding not only the possibility that markets may *fail* in certain instances but also the offer of guidance on how they may then be *fixed*. In particular, this then enables a resurgent argument in favor of (possibly "strong") states rectifying market failures, perhaps through strong regulatory intervention and public ownership of infrastructure. Certainly, the government and the state itself thereby receive a renewed legitimacy and mission, in terms of being the agency responsible for managing the stability and resilience of the "system" (at whatever territorial scale) as a whole. Clearly, this could readily enable a renewed and newly legitimate project of public investment in energy and environment, and innovation and infrastructure related thereto, that has been a singular and deliberate absence throughout the neoliberal period.

What is most important about this putative future regime, however, is that this is not merely an abstract speculative possibility but rather one that emerges from a reading of the ongoing evolution of sociotechnical and political change within the presumptive heir apparent for global capitalist hegemony—namely, China.[45] While many pieces of the puzzle remain embryonic or even absent, such is the dynamism in China across the gamut of issues related to questions of the parallel development of political-economic regimes (national, local, and indeed, global) and innovation—and especially regarding issues of energy and environment and particularly in the key space of emerging digital innovations in both of these domains—that betting *against* significant qualitative change in Chinese society and politics in the medium term seems almost the bigger risk.

Moreover, many of the key elements of this emergent regime are clearly in place and fast developing. For instance, in contemporary China, we find a singularly dynamic participative—if pragmatic—social media–based public emerging, concentrated in a buoyant and rising urban "middle class" and its production-side equivalent of a surging private sector economy, including in digital and knowledge economy service sectors. This group also takes environmental risks (increasingly including climate change) extremely seriously as day-to-day lived matters of ontological security.

Such concerns, however, and their empowerment with the ascendancy of this power bloc on both national and global stages do not augur a new universalistic concern with "saving the planet" in China. Rather, the parallel and interdependent emergence of both concerted efforts on such issues and this particular constituency most likely suggest that dealing with environmental challenges will unfold in ways that systematically privilege the concerns of this emergent sociopolitical constituency while neglecting the more numerous majority who remain disproportionately exposed to the risks and dangers of environmental change.

This is thus a systematically duplicitous regime in which a gloss of environmental and entrepreneurial virtue enables and is enabled by continuing socioeconomic inequalities—again, just as the nineteenth-century gospel of liberal "progress" was coproduced with the development of the regime of industrial capitalism and its unprecedented Dickensian inequalities at home and creeping imperialism abroad. Symptomatic here would be the growth of a narrative of China as "green savior," highlighting its world-leading environmental regulation and investment in renewables while occluding its continued reliance on and investment in fossil fuels, including coal, both domestically and, through its massive investment program of "One Belt, One Road," overseas.[46]

In this scenario, then, at least a credible "patch" of climate engineering technology would be needed as promise and deployment, while China's already-established leadership in production of renewable energy technologies would suggest climate engineering may also become strongly advocated and pursued without reference to fossil fuels. The current remaining interest in fossil CCS and the growing interest in (some) CDRs may thus here be taken up[47] and the technologies developed and implemented to some extent driven by regulation. SRM seems a more distant prospect—but also not inconceivable.

On the one hand, with little interest or expertise in China on SRM evident to date[48] and with the singular sensitivity to the Chinese government of issues of national sovereignty—issues raised by SRM—there is evidence that China will be reluctant to pursue this technology, seeing it as too dangerous and prone to lead to international conflict.[49] Though more of a stretch, some have also argued that Chinese policy and diplomacy are structurally and/or culturally disinclined either to take the lead in such a hugely controversial global experiment or to play with the "human-nature" balance with such seeming recklessness. Yet some of these same reasons also suggest that, while not taking a unilateral lead in pursuing SRM, China could emerge as a crucial mediator and negotiator in constructing global experiments and regulations that allow such development and deployment to take place.

Here, in other words, by stressing the *break* with neoliberal U.S. unilateral dominance that it represents—manifest in both its approach to regulation of climate engineering and its "leadership" on climate matters and "green technologies"—China may yet construct new political feedback loops in which development of SRM, with China as *primus inter pares*, serves to boost its global (and domestic) political standing precisely as "green savior" and responsible global custodian. In this scenario, then, climate engineering (including SRM) may be developed in ways that explicitly distance these technologies from the case for continued fossil fuel use with which they were associated under neoliberalism—perhaps helpfully occluding the continued investment in fossil fuel infrastructures by China in the meantime. As the climate crisis deepens, the uptake of climate engineering as discussed previously becomes more likely, especially if the impacts are felt strongly in China.

The Contemporary Hinge

The three future scenarios outlined previously extrapolate observable tendencies in the current situation. Taken together, the scenarios show that the current moment holds the potential for coevolution of the political regime and climate engineering into different possible futures. This is not surprising, but we can use the specifics of the scenarios to shed light on the structure of the *current* turbulence by exploring the relations between the observed tendencies. We are not arguing that the scenarios are the given trajectories, the paths that must be chosen among, in a predictive mode analysis—and so we are not seeking to identify choice points. Rather, we discuss relations between scenarios—and the capital fractions dominant in each—as a way of identifying important dialectical tensions in the current moment and their impact on climate engineering. We use the continued neoliberal scenario as a default or baseline and discuss its relations with the two other scenarios presented previously.

Neoliberalism—Illiberalism

As discussed previously, Trump (alongside Brexit, Italy's populist coalition, Alternativ für Deutschland, Le Pen, etc.) can be read as an (immanent) illiberal challenge to the neoliberal regime, but the latter has not simply disappeared or disintegrated. We still see a strongly financialized economy—fueled by growth in consumer debt[50] and with Wall Street influence thus undiminished—that is a globalized financial capital fraction sustaining U.S. hegemony. The neoliberal regime has also seen the birth of and been sustained by strong growth in high-tech giants in Silicon Valley and beyond, getting credit from Wall Street and generating a new generation of tech billionaires,[51] albeit one still thoroughly saturated with the mores, expectations, and power relations of neoliberal (venture) finance. The underpinning server infrastructure is fueled mainly by fossil fuels—lately, increasingly with natural gas. The neoliberal regime has shaped the policies of both the main political parties in the United States. However, we now also see some industry support for the Trump administration, with its often illiberal policies.[52] This is no doubt in part driven by a hope for further deregulation, tax cuts, and privatization—for example, in the climate policy arena. We can thus see a tension between neoliberal and illiberal futures.

Neoliberal climate policy has been shaped in the tension between climate denialism and climate activism (alongside nearly all climate science). While denialism has a far stronger presence in the Trump administration than it did in the Clinton and Obama ones, and in this sense an improved standing, there is still a broad coalition of actors in support of climate policy, including states, civil society organizations and parts of industry,[53] and international pressure (especially after the Paris Agreement).

Trump campaigned on leaving the UNFCCC Paris Agreement, but the administration later wavered on this issue. Trump's aides were split on the issue, as is the Republican Party.[54] Even some big coal companies argued for staying in, with a central argument being to retain a seat at the international climate policy negotiation table (and that their investment planning horizons are longer than a presidency).[55] A strategy of staying in but doing very little was a possible outcome[56] and would have been in line with neoliberal superficial climate policy, even though the Trump administration finally decided to withdraw the United States from the Paris Agreement.

Uncertainty over U.S. climate policy is affecting the perceived prospects of climate engineering. There is speculation that if Trump can't fight off climate policy as an issue, the potentially cheap fix of SRM may start looking attractive to the administration[57] and likely rather more attractive than expensive CCS facilities that would add to costs of using coal (and gas). Prominent climate engineering researchers have stated, however, that they would not take funding from the administration and would counsel others to do likewise, with the argument to avoid the technology being associated with the denialist elements of the administration[58] (although others would say that is too late[59]).

But high-profile climate scientists have also warned that Trump rolling back climate change policy may further undermine belief in the adequacy of current mitigation-oriented climate policies and in turn strengthen the case for (at least researching) climate engineering,[60] which would then have to be funded by others. It is worth noting that the Gates Foundation will likely fund a new Harvard-based research program on climate engineering,[61] alongside other foundations. An op-ed in the San Diego Union-Tribute went so far as to say, "We call on one or more of the nation's benevolent billionaires to consider privately funding a massive endeavour on the scale of the Manhattan Project to try to geoengineer such a hedge."[62]

Key industries of the neoliberal regime, on Wall Street and beyond, may opt to support (or at least research) climate engineering independently of the federal government. This has already happened but may now ironically get a boost by fears of illiberal denialism (justified with reference to the coal industry among others). If, on the other hand, the Trump administration were forced to side with neoliberalism and rejuvenate its climate policy, then it may be tempted to support climate engineering research as a useful promise of a cheap, apparently nondisruptive technical fix. We may even see parallel competing efforts of the administration and private actors.

The current tension between neoliberalism and illiberalism creates new uncertainty for climate policy, but climate engineering may come out of it rather well. The promise of climate engineering may be useful for powerful actors, both neoliberal and populist, as long as these tensions remain, and support for climate engineering research may reap the benefits of this.

Neoliberalism—Liberalism 2.0

China has surpassed the United States in terms of trade volume but not yet GDP and has as yet not the same global reach in terms of military, political, or cultural influence. Clearly though, there is a geopolitical tension between the countries. Importantly, as set out previously, a new Chinese hegemony might take the form not of neoliberalism (or illiberalism) but of liberalism 2.0.[63] The geopolitical tension is thus also a tension between political regimes.

China appears to be taking climate change increasingly seriously. This is due in part to rising domestic concerns about pollution and environmental harms and in support of a booming cleantech industry, but it is also a way to project soft power internationally.[64] We should remember though that the Chinese economy is strongly coal-dependent and that China exports large numbers of coal plants to other countries especially in its Asian regional neighborhood,[65] which might mean that China will not take a global leadership role on climate policy.[66]

This uncertainty over the relative positions of the United States and China globally, in general, and regarding climate policy, in particular, has probably not had much effect on climate engineering prospects so far. But that could now be changing as Trump pulled the United States out of the Paris Agreement, giving China an even better opportunity to challenge the United States for climate policy leadership[67] while President Xi has made repeated and explicit global statements to bolster China's image as responsible global custodian. China might also want to exploit the rift between the United States and Europe on climate policy (and possibly now even national security) and strengthen its collaboration with Europe on climate policy in order to bolster its own green credentials and at the same time tie the continent closer economically and politically.

While there is a track record in China of weather modification[68] and a recent national research program on climate engineering[69] and CCS research and demonstration, there seems to be little indication of support for SRM deployment. Hamilton has argued that climate stress could make Chinese people desperate and that a government under pressure might be tempted to deploy cheap SRM in response.[70] However, several commentators see minimal likelihood of Chinese unilateral SRM deployment due to the lack of any significant constituency advocating for it coupled with the incompatibility of SRM deployment with China's strong protection of sovereignty as a principle in international relations.[71]

If China wants to demonstrate its commitment to climate mitigation, increased support for CCS or CDR technologies seems more likely. This would also seem to be most straightforwardly compatible with a future liberalism 2.0 in terms of a somewhat stronger reliance on governmental nature stewardship and regulation limiting the worst excesses of markets and private action and through action by (networks of) private business and/or civil society actors (e.g., energy companies doing CCS or activists and NGOs [nongovernmental organizations] running afforestation projects, albeit perhaps in both cases in the context of considerable governmental carrots and sticks). Such a model of fiercely competitive and experimental private enterprise taking the lead in constructing new technological capabilities and sectors on the basis of even vague and uncertain policy promises of the uniquely massive support of the Chinese central government is certainly already evident in other cognate sectors such as solar thermal or concentrated solar power.[72] Conversely, at least insofar as it is imagined as a unilateral and radically self-interested technical fix, such complexity liberalism would seem to fit poorly with SRM deployment.

If China continues to challenge the United States in terms of global climate policy leadership as part of a wider challenge for global influence, it will become increasingly difficult for U.S. denialists to roll back U.S. climate commitments internationally and domestically. For U.S. neoliberals, this is an argument for continued market-based climate policy propped up by CCS and CDR promises, legitimized by research. For U.S. illiberals, this situation might make climate engineering promises more appealing too, and they may support SRM more willingly.

In the tension between a continued, but threatened, neoliberal regime and a budding liberalism 2.0, we can also see a dynamic that might benefit climate engineering (possibly framed *as against* fossil fuel use). In the short term, we might expect to see research and maybe even some deployment as an embryonic liberal 2.0 regime logic argues, perhaps not unpersuasively, for the need to explore "all options" to reduce the potentially catastrophic effects of global warming as a position of responsibility and responsiveness to the climate emergency, not its denial or technical fix. This would thus involve rhetorical strategies to distance development of climate engineering from programs that seek to deploy it so as to either allow unmitigated growth of fossil fuel consumption (as per the illiberal scenario) or buy indefinite time for the market to solve the challenge of emissions and to enable marketization of the climate (as in the neoliberal one). Instead, a liberal 2.0 climate engineering would seek to wed itself firmly in the public imagination with tackling climate change and decarbonization, and with doing so "responsibly," it would hence balance the dangers of developing and not developing climate engineering capacities, deploying and regulating it multilaterally, and so on. For now, this approach would likely benefit CCS and CDR most, but SRM may also emerge, reframed as explicitly constrained (as discussed previously) and as a focus of such "responsible" experimentation.[73]

Finally, perhaps bringing all three scenarios together, legitimacy of experimentation and advancing these technological options must be considered.[74] A new language and promise of "responsible" experimentation with climate engineering may well be motivated precisely by the worsening crisis of populism, its environmental irresponsibility, its political-economic incompetence, and the social, political, economic, and financial instability it fuels. Here, then, such a dialectical inversion would suggest that development along the illiberal trajectory may fuel the emergence of the China-centered liberal 2.0 scenario, but not vice versa. For the latter would move progressively toward meaningful action on climate change (benefitting at least a powerful "some") that would serve to release some of the dissatisfaction fueling the populist surge, while the essentially nihilistic force of populism can lead only to further grievance, division, and self-destruction.

Conclusion

This chapter has explored the impact and significance of Trump (as an indicator of and shorthand for a wider process of geopolitical change in the form of a chauvinistic populism of political-economic retrenchment) regarding the coevolution of political regimes with promises of technical fixes to the problem of global climate change. Three radically different but possible future scenarios were described, and tensions among possible futures created a lens for interpreting the current moment. We identified two key dialectics: between neoliberalism and illiberalism and between continued neoliberal (but illiberally challenged) U.S. hegemony and budding China-centered liberalism 2.0 (and the dominant fractions of capital corresponding to each of these regimes). Both these dialectics appear conducive to prolonged attention to the climate engineering promise, whether as talk, research, or limited deployment. In a nutshell, while continued neoliberalism in the United States and elsewhere may lead to government support for cheap-seeming climate engineering, illiberal denialism may boost nonstate neoliberal efforts to support climate engineering. And while China-based liberalism 2.0 might lead to "responsible" climate engineering efforts, even a continued U.S. hegemony will be challenged by Chinese climate concern enough to justify United States support for climate engineering as a fix.

In our previous paper,[75] we analyzed how the multiplicity of climate engineering as a promise of a technical fix to climate change has mattered for its coevolution with the neoliberal political economy regime. In this chapter, with its prospectively oriented analysis, we also discussed how a multiplicity of possible political economy regimes matters for that coevolution. We were prompted to think beyond just a continued unfolding of the neoliberal regime by the election of Trump, and we developed a method of systematically exploring our current predicament by juxtaposing different scenarios.

Acknowledging turbulence in the current moment, we have sought to analyze emergent structures within that turbulence. This exploration does not presume scope for prediction; rather when we compare analysis from a few years ago,[76] we reflect on what recent events might mean for the possible coevolutions of the political regime with climate engineering promises. Within the last few years, a newly realized illiberal challenge to the neoliberal regime has altered the prospects and pathways for climate engineering. The *relative* likelihood of the illiberal scenario versus continued neoliberalism perhaps seems to have increased in the short term. Although, taking dialectical inversions and the turbulence of populist politics that are built into the former, this could be dramatically rebalanced in the opposite direction—for example, given unforeseeable (possibly "black swan") political "events" that make Republican-led congressional impeachment of Trump unavoidable. In any case, before we start mourning neoliberalism, it is important to acknowledge its insufficiency for promoting climate change action, at least at the scale and pace needed given the objective extent of this planetary emergency. Neoliberalism also undermined the very concept of a "public reason"—namely, debate in the public sphere about matters of public good and accepted as, or at least held to account as, rational and based on empirical facts, favoring markets as epistemic arbiters. Trumpian post-truth is another aspect of how the populism for which he is the current figurehead is the child of neoliberalism, and neoliberalism has now indelibly conditioned whatever comes next—whether it be a renewed invigorated neoliberalism or a new illiberal regime with autocratic rule.

Recognizing that we are currently living through a time of instability and crisis, some may assume illiberalism may be imminent. This moment is, however, unique and unpredictable. Comparing this time to the 1930s crisis and the rise of fascism in Europe, significant differences must be acknowledged. For starters, it should be noted that at least in the United States, it is the opponents, not supporters, of illiberal insurgency that currently master the mainstream media and that are on the streets en masse, whether in marches for climate, science, or women. While excruciatingly frustrating and certainly not without significant future jeopardy, the United States at the time of writing continues to resist attempts to dismantle the rule of law and constitutional government in ways that Germany in 1933 failed to. But another big difference concerns the deepening capitalist embrace and the sheer economic strength of China. In the area of climate policy, as we have argued previously, a Trump regime rollback of U.S. climate policy positions is also an opening for Chinese global positioning that the leadership of the Chinese Communist Party seems to be grasping with two hands.

With the coevolution of climate engineering advancement and political regimes, in the short term, we might continue to expect limited research and little implementation of CCS, CDR, or SRM. We expect policy debate about CDR—especially BECCS—to continue, since many experts assume it is needed to halt the most dangerous climate scenarios that may provoke panic.

However, the illiberal challenge to neoliberalism has now changed the game, and this exploratory consideration suggests that there may now be *more* support for climate engineering research rather than less. The illiberal boost of denialism is likely to spark *growing* attention to SRM, and we can expect ongoing controversy over this class of technologies in domestic and international discourse. Insofar as there is increased experimentation and deployment of climate engineering technologies, a variety of different forms of backlash might be expected from diverse political agencies and positions and with uncertain effects. Controversy regarding climate engineering as a promise of a technical fix to climate change is unlikely to disappear as long as the climate problem continues to worsen, as long as climate activists continue to resist, and as long as fossil interests are central to political regimes. Of course, the irony here is that this dynamic and resilient tumult of forces arraigned around the issue of climate engineering may also likely shape such developments as they happen in ways that could well make it increasingly palatable, and even positively attractive, for the great many who do not engage in any depth in climate engineering issues and politics, even as it remains always essentially contested—in ways that clearly resonate with the unfolding of a liberalism 2.0.

Of course, other scenarios beyond the three articulated earlier are also possible. Previously, we explored a fourth scenario with a strongly participative, radically democratic climate engineering with bottom-up, local engagement with the climate. This scenario would require a rearticulation of both climate engineering and the climate change problem and a focus on relatively low-tech and small-scale technologies. Martindale analyses Transition Town practices as potential democratic climate engineering,[77] and here we have moved beyond what would currently be labeled as technical fixes, as such practices are typically not readily quantified and commodified.[78] This kind of low-tech, localist "climate engineering" would require a radically different political regime—with small-scale, local orientation governed by a town hall–scale public reason—and would likely require definitive moves beyond capitalism to avoid centralization and up-scaling and enable local, communal forms of ownership to dominate. This scenario seems highly unlikely on the time scales we have discussed here.

There are, however, also other progressive, if less utopian, visions in the literature. Olson[79] sets out criteria for geoengineering variants that "touch gently on biological and social systems," leading him to focus on some technologies that can scale from the local to the regional. Similarly, Buck's[80] focus on participation leads her to focus on regional terrestrial climate engineering technologies, in explicit contrast to what globally central actors can do top-down for the world as a whole. These latter versions are less uncompromisingly utopian, plausible within a still-capitalist system and therefore more easily promoted as elements of a progressive politics of resistance against, engagement with, and socioecological transformation of the political regimes discussed previously. There is an urgent need for further analysis of the potential role of socially progressive politics in the coevolution of climate engineering and political economies.

Notes

1 Carol J. Clouse, "Donald Trump Supports 'Clean Coal'—but Does It Really Have a Future?," *Guardian*, December 4, 2017, http://www.theguardian.com/sustainable -business/2016/dec/04/trump-supports-clean-coal-but-does-it-really-have-a-future; John Schwartz, "Can Carbon Capture Technology Prosper under Trump?," *New York Times*, January 2, 2017, http://www.nytimes.com/2017/01/02/science/donald-trump-carbon -capture-clean-coal.html; Faber et al., "Trump's Electoral Triumph: Class, Race, Gender, and the Hegemony of the Polluter-Industrial Complex," *Capitalism Nature Socialism* 28, no. 1 (2017): 1–15.

2 Nils Markusson et al., "The Political Economy of Technical Fixes: The (Mis)Alignment of Clean Fossil and Political Regimes," *Energy Research & Social Science* 23 (2017): 1–10.

3 Larry Lohmann, "Financialization, Commodification and Carbon: The Contradictions of Neoliberal Climate Policy," *Socialist register* 48, no. 85 (2012): 107; John Quiggin, David Adamson, and Daniel Quiggin, eds., *Carbon Pricing: Early Experience and Future Prospects* (Cheltenham, Pa.: Edward Elgar, 2014), 183.

4 James McDougall, "No, This Isn't the 1930s—but Yes, This Is Fascism," The Conversation, November 16, 2016, http://theconversation.com/no-this-isnt-the-1930s-but-yes-this-is -fascism-68867.

5 Martin Lukacs, "Trump Presidency 'Opens Door' to Planet-Hacking Geoengineer Experiments," *Guardian*, March 27, 2017, http://www.theguardian.com/environment/ true-north/2017/mar/27/trump-presidency-opens-door-to-planet-hacking-geoengineer -experiments; Aleszu Bajak, "The Dangerous Belief That Extreme Technology Will Fix Climate Change," *Huffington Post*, April 27, 2018, http://www.huffpost.com/entry/ geoengineering-climate-change_n_5ae07919e4b061c0bfa3e794.

6 Phil Macnaghten and Bronislaw Szerszynski, "Living the Global Social Experiment: An Analysis of Public Discourse on Solar Radiation Management and Its Implications for Governance," *Global Environmental Change* 23, no. 2 (2013): 465–474.

7 Immanuel Wallerstein, *The Decline of American Power: The U.S. in a Chaotic World* (New York: New Press, 2003), 324.

8 David W. Keith and Gernot Wagner, "Fear of Solar Geoengineering Is Healthy—but Don't Distort Our Research," *Guardian*, March 29, 2017, http://www.theguardian.com/ environment/2017/mar/29/criticism-harvard-solar-geoengineering-research-distorted.

9 Colin Crouch, *The Strange Non-death of Neoliberalism* (Cambridge, U.K.: Polity, 2011), 212.

10 Christian Fuchs, "Donald Trump: A Critical Theory-Perspective on Authoritarian Capitalism," *tripleC: Communication, Capitalism & Critique* 15, no. 1 (2017): 1–72.

11 David Smith, "Trump Begins Tearing Up Obama's Years of Progress on Tackling Climate Change," *Guardian*, March 28, 2017, http://www.theguardian.com/us-news/2017/mar/ 28/trump-begins-tearing-up-obamas-years-of-progress-on-tackling-climate-change; Valerie Volcovici, Nichola Groom, and Scott DiSavino, "Trump Declares End to 'War on Coal,' but Utilities Aren't Listening," *Reuters*, April 5, 2017, http://www.reuters.com/ article/us-usa-trump-climate-power/trump-declares-end-to-war-on-coal-but-utilities -arent-listening-idUSKBN1770D8.

12 Lukacs, "Trump Presidency 'Opens Door'"; Keith and Wagner, "Fear of Solar Geoengineering."

13 Markusson et al., "Political Economy of Technical Fixes."

14 Marianne Ryghaug and Thomas Moe Skjølsvold, "The Global Warming of Climate Science: Climategate and the Construction of Scientific Facts," *International Studies in the Philosophy of Science* 24, no. 3 (2010): 287–307.

15 Markusson et al., "Political Economy of Technical Fixes."

16 Philip Mirowski, *Never Let a Serious Crisis Go to Waste: How Neoliberalism Survived the Financial Meltdown* (London: Verso, 2013), 480.

17 David Tyfield, "Science, Innovation and Neoliberalism," in *"Science, Innovation and Neo-liberalism": The Routledge Handbook of Neoliberalism*, ed. Simon Springer, Kean Birch, and Julie Macleavy (London: Routledge, 2016), chap. 29.

18 Timothy Mitchell, *Carbon Democracy: Political Power in the Age of Oil* (London: Verso, 2011), 288.

19 Jennie C. Stephens, "Time to Stop CCS Investments and End Government Subsidies of Fossil Fuels," *Wiley Interdisciplinary Reviews: Climate Change* 5, no. 2 (2014): 169–173; Jennie C. Stephens, "Carbon Capture and Storage: A Controversial Climate Mitigation Approach," *International Spectator* 50, no. 1 (2015): 74–84.

20 Markusson et al., "Political Economy of Technical Fixes."

21 Jennie C. Stephens, "Technology Leader, Policy Laggard: Carbon Capture and Storage (CCS) Development for Climate Mitigation in the U.S. Political Context," in *Caching the Carbon: The Politics and Policy of Carbon Capture and Storage*, ed. James Meadowcroft and Oluf Langhelle (Cheltenham: Edward Elgar, 2009), 22–49.

22 Nils Markusson, Atsushi Ishii, and Jennie C. Stephens, "The Social and Political Complexities of Learning in CCS Demonstration Projects," *Global Environmental Change* 21, no. 2 (2011): 293–302.

23 Markusson et al., "Political Economy of Technical Fixes."

24 Benjamin Evar, "Framing CO_2 Storage Risk: A Cultural Theory Perspective," *Energy & Environment* 23, nos. 2–3 (2012): 375–387; Alfonso Martinez Arranz, "Hype among Low-Carbon Technologies: Carbon Capture and Storage in Comparison," *Global Environmental Change* 41 (2016): 124–141.

25 The Royal Society, *Geoengineering the Climate: Science, Governance and Uncertainty* (London: Royal Society, 2009), 89; International Panel on Climate Change (IPCC), *Climate Change 2014: Synthesis Report; Contribution of Working Groups I, II and III to the Fifth Assessment Report of the Intergovernmental Panel on Climate Change* (Geneva, Switzerland: IPCC, 2014), 151.

26 See discussion in Surprise, this volume, chap. 13.

27 "Some Thoughts on the Crisis of Liberalism—and How to Fix It," Bagehot's Notebook, *Economist*, June 12, 2018, http://www.economist.com/bagehots-notebook/2018/06/12/some-thoughts-on-the-crisis-of-liberalism-and-how-to-fix-it.

28 Stephen R. Gill and David Law, "Global Hegemony and the Structural Power of Capital," *International Studies Quarterly* 33, no. 4 (1989): 475–499; William I. Robinson and Jerry Harris, "Towards a Global Ruling Class? Globalization and the Transnational Capitalist Class," *Science and Society* 64, no. 1 (2000): 11–54.

29 Giovanni Arrighi, *The Long Twentieth Century: Money, Power and the Origins of Our Times* (London: Verso, 2010), 432.

30 Faber et al., "Trump's Electoral Triumph."

31 Will Hutton, "Instead of Draining the Swamp, Trump Has Become Wall Street's Best Buddy," *Guardian*, February 12, 2017, http://www.theguardian.com/commentisfree/2017/feb/12/not-draining-swamp-trump-money-mens-best-friend; Thomas Palley, "Trumponomics: Neocon Neoliberalism Camouflaged with Anti-globalization Circus," Social Europe, April 17, 2017, http://www.socialeurope.eu/2017/04/trumps-international-economic-policy-neocon-neoliberalism-camouflaged-anti-globalization-circus/.

32 Bob Jessop, "The Organic Crisis of the British State: Putting Brexit in Its Place," *Globalizations* 14, no. 1 (2016): 133–141; David Keith, "Toward a Responsible Solar Geoengineering Research Program," *Issues in Science and Technology* 33, no. 3 (2017).

33 Mitchell, *Carbon Democracy*, 288; David Tyfield, "'King Coal Is Dead! Long Live the King!': The Paradoxes of Coal's Resurgence in the Emergence of Global Low-Carbon Societies," *Theory, Culture & Society* 31, no. 5 (2014): 59–81.

34 McDougall, "This Isn't the 1930s"; Faber et al., "Trump's Electoral Triumph."

35 Crouch, *Non-death of Neoliberalism*, 212.

36   Scott Smith and Georgina Voss, "Will Populism Kill Your Jetpack? What Trump and Brexit Mean for Self-Driving Cars, Renewable Energy and Future Breakthroughs," *Atlantic*, December 16, 2016, http://www.theatlantic.com/technology/archive/2016/12/will-populism-kill-your-jetpack/510734/.

37   Patti Waldmeir, "Donald Trump Opens $10bn Foxconn Factory in Wisconsin," *Financial Times*, June 29, 2018, http://www.ft.com/content/73b63d08-7b08-11e8-bc55-50daf11b720d.

38   Clouse, "Donald Trump Supports 'Clean Coal'"; Schwartz, "Can Technology Prosper?"

39   Macnaghten and Szerszynski, "Living the Global Social Experiment," 465–474.

40   David G. Victor, "On the Regulation of Geoengineering," *Oxford Review of Economic Policy* 24, no. 2 (2008): 322–336.

41   James Rodger Fleming, *Fixing the Sky: The Checkered History of Weather and Climate Control* (New York: Columbia University Press, 2010), 344.

42   See Surprise, this volume, chap. 13.

43   David Tyfield, *Liberalism 2.0 and the Rise of China: Global Crisis, Innovation, Urban Mobility* (London: Routledge, 2017), 264.

44   Tyfield, "King Coal Is Dead!"; Tyfield, *Liberalism 2.0*, 264.

45   *Tyfield, "King Coal Is Dead!"; Tyfield, Liberalism 2.0, 264.*

46   Beth Walker, "China Stokes Global Coal Growth," *Chinadialogue*, September 23, 2016, http://www.chinadialogue.net/article/show/single/en/9264-China-stokes-global-coal-growth; Benjamin Haas, "Climate Change: China Calls US 'Selfish' after Trump Seeks to Bring Back Coal," *Guardian*, March 30, 2017, http://www.theguardian.com/world/2017/mar/30/climate-change-china-us-selfish-trump-coal; Feng Hao, "China's Belt and Road Still Pushing Coal," *Chinadialogue*, May 12, 2017, http://www.chinadialogue.net/article/show/single/en/9785-China-s-Belt-and-Road-Initiative-still-pushing-coal.

47   Weili Weng and Ying Chen, "A Chinese Perspective on Solar Geoengineering," in *Geoengineering Our Climate, A Working Paper Series on the Ethics, Politics and Governance of Climate Engineering*, ed. Jason J. Blackstock and Sean Low (London: Routledge, 2018), chap. 23.

48   James Temple, "China Builds One of the World's Largest Geoengineering Research Programs," *MIT Technology Review*, August 2, 2017, http://www.technologyreview.com/s/608401/china-builds-one-of-the-worlds-largest-geoengineering-research-programs/.

49   John C. Moore et al., "Will China Be the First to Initiate Climate Engineering?," *Earth's Future* 4, no. 12 (2016): 588–595.

50   U.S. Federal Reserve, *Quarterly Report on Household Debt and Credit* (New York: U.S. Federal Reserve, May 2017.)

51   Smith and Voss, "Will Populism Kill Your Jetpack?"; "Apple Should Shrink Its Finance Arm before It Goes Bananas," *Economist*, October 28, 2017, http://www.economist.com/business/2017/10/28/apple-should-shrink-its-finance-arm-before-it-goes-bananas.

52   "Why Corporate America Loves Donald Trump," *Economist*, May 24, 2018, http://www.economist.com/leaders/2018/05/24/why-corporate-america-loves-donald-trump.

53   J. P. Sapinski, "Climate Capitalism and the Global Corporate Elite Network," *Environmental Sociology* 1, no. 4 (2015): 268–279; J. P. Sapinski, "Corporate Climate Policy-Planning in the Global Polity: A Network Analysis," *Critical Sociology* 45, nos. 4–5 (2017): 565–582; Smith, "Trump Begins Tearing Up."

54   Oliver Milman, "Trump Aides Abruptly Postpone Meeting on Whether to Stay in Pairs Climate Deal," *Guardian*, April 18, 2017, http://www.theguardian.com/us-news/2017/may/09/paris-climate-deal-trump-advisers.

55   Valerie Volcovici, "U.S. Coal Companies Ask Trump to Stick with Paris Climate Deal," *Reuters*, April 4, 2017, http://www.reuters.com/article/us-usa-trump-coal-idUSKBN1762YY.

56   Milman, "Trump Aides."

57 Eduardo Porter, "To Curb Global Warming, Science Fiction May Become Fact," *New York Times*, April 4, 2017, http://www.nytimes.com/2017/04/04/business/economy/geoengineering-climate-change.html; Bajak, "Dangerous Belief."

58 Keith and Wagner, "Fear of Solar Geoengineering"; Keith, "Responsible Solar Geoengineering."

59 ETC Group, BiofuelWatch, and Heinrich Böll Stiftung, *The Big Bad Fix: A Case against Geoengineering* (Val David, Canada: ETC Group, 2017).

60 Arthur Neslen, "US Scientists Launch World's Biggest Solar Geoengineering Study," *Guardian*, March 24, 2017, http://www.theguardian.com/environment/2017/mar/24/us-scientists-launch-worlds-biggest-solar-geoengineering-study; Porter, "To Curb Global Warming."

61 Keith and Wagner, "Fear of Solar Geoengineering"; Neslen, "World's Biggest Solar Geo-engineering Study."

62 San Diego Union-Tribune, "Climate Change: Trump's Dramatic Move Deserves Dramatic Response," *San Diego Union-Tribune*, March 17, 2017, http://www.sandiegouniontribune.com/opinion/editorials/sd-climate-change-geoengineering-trump-20170316-story.html.

63 Tyfield, "King Coal Is Dead!"; Tyfield, *Liberalism 2.0*, 264.

64 David Tyfield, "Why China Is Serious about Becoming the Global Leader on Climate Change," The Conversation, April 10, 2017, http://theconversation.com/why-china-is-serious-about-becoming-the-global-leader-on-climate-change-75762.

65 Walker, "China Stokes Global Coal Growth."

66 Haas, "Climate Change."

67 Wu Wenyuan, "China's 'Clean Coal' Power: A Viable Model or Cautionary Tale?," *Chinadialogue*, June 27, 2017, http://www.chinadialogue.net/blog/9876-China-s-clean-coal-power-A-viable-model-or-cautionary-tale-/en.

68 Xueliang Guo and Guoguang Zheng, "Advances in Weather Modification from 1997 to 2007 in China," *Advances in Atmospheric Sciences* 26, no. 2 (2009): 240–252; Kevin Lui, "China Is Splashing $168 Million to Make It Rain," *Fortune*, January 24, 2017, http://fortune.com/2017/01/24/china-government-artificial-rain-program/.

69 Long Cao, Chao-Chao Gao, and Li-Yun Zhao, "Geoengineering: Basic Science and Ongoing Research Efforts in China," *Advances in Climate Change Research* 6, nos. 3–4 (2015): 188–196.

70 Clive Hamilton, *Earthmasters: The Dawn of the Age of Geoengineering* (New Haven, Conn.: Yale University Press, 2013).

71 Kingsley Edney and Jonathan Symons, "China and the Blunt Temptations of Geo-engineering: The Role of Solar Radiation Management in China's Strategic Response to Climate Change," *Pacific Review* 27, no. 3 (2013): 307–332; Weng and Chen, "Chinese Perspective on Solar Geoengineering."

72 Jorrit Gosens, "Mechanisms of Unrelated Diversification Enabling China's Demonstration of Nascent Renewable Energy Technologies: The Case of Concentrated Solar Power" (paper presented at the International Sustainability Transitions [IST] conference, Manchester, June 11–14, 2018).

73 Keith, "Responsible Solar Geoengineering"; Sean Low, "Engineering Imaginaries: Anticipatory Foresight for Solar Radiation Management Governance," *Science of the Total Environment* 580 (2016): 90–104.

74 Peter C. Frumhoff and Jennie C. Stephens, "Toward Legitimacy in the Solar Geoengineering Research Enterprise," *Philosophical Transactions of the Royal Society A* 376, no. 2119 (2018).

75 Markusson et al., "Political Economy of Technical Fixes."

76 Markusson et al.

77   Leigh Martindale, "Understanding Humans in the Anthropocene: Finding Answers in Geoengineering and Transition Towns," *Environment and Planning D: Society and Space* 33, no. 5 (2015): 907–924.

78   Larry Lohmann, "Marketing and Making Carbon Dumps: Commodification, Calculation and Counterfactuals in Climate Change Mitigation," *Science as Culture* 14, no. 3 (2005): 203–235.

79   Robert L. Olson, "Soft Geoengineering: A Gentler Approach to Addressing Climate Change," *Environment: Science and Policy for Sustainable Development* 54, no. 5 (2012): 30.

80   Holly Jean Buck, "Geoengineering: Re-making Climate for Profit or Humanitarian Intervention?," *Development and Change* 43, no. 1 (2012): 253–270.

15

Prospects of Climate Engineering in a Post-truth Era

HOLLY JEAN BUCK

For too long, geoengineering has been treated as another technology coming down the pipes as we're thrust into a brave new century—a new thing that's inevitably on its way, like self-driving cars or blockchain or designer babies, except there's no opt-in or opt-out. Of course, geoengineering would be a program, an intervention that happens through time and is shaped by its context. Yet most analysts and journalists continue to treat it with implicit technological determinism. When the context is assumed, it is still assumed to be one of civil geopolitics and rational actors upon a backdrop of U.N. corridors or the bland glass facades of the lobbyists on K Street. "Decision-makers" are unspecified but assumed to exist.

And so much of the writing and speculation on geoengineering proceeds as if it is still the 1990s outside. Perhaps commentators don't know what else to do, since the state we're now in is still pretty murky. The easiest thing is to continue as if there has been no significant break from the norms, values, and structures of the post–World War II era. As if government as usual is proceeding within standard parameters. In the United States, though, the credibility of this rational facade of normal decision-making is being stretched to the breaking point by the blatant lies of the Trump administration in turn enabled by Republican lawmakers. Social currents are stirring up the violence and inequality beneath the surface of decision-making: the rise of the alt-right and the #MeToo movement both surface what lies beneath in different ways. It is harder and harder to mask the troubled context—and this is crucial for thinking about the prospects of geoengineering.

While there are certain risks inherent in some climate engineering technologies, much of their risks actually derives from the context in which these strategies are researched, developed, and promoted. The election of Donald Trump as president of the United States is the latest, and probably most significant, disruption of the context in which geoengineering may emerge. Other countries are not immune to strongmen and underlying social strife, and the ills are clearly not unique to the United States. But they manifest in a particular way in the United States and are notably concerning there because of the power the United States still wields in global affairs. As the United States possesses a great deal of technological capacity for climate engineering, this chapter will focus on dimensions of the U.S. context—chiefly, the status of truth. It may not be possible for any credible or rational form of geoengineering to emerge in this context—and if not, then what?

Truth has been a central concern in U.S. politics since the founding of the republic. Jill Lepore, in her seminal political history *These Truths*, points to Thomas Paine's "simple facts, plain arguments, and common sense," which represented both a new method of thinking and a new era of politics, in Paine's words. The nation was based on Enlightenment values of inquiry, reason, and facts but was always in tension, since it couldn't reconcile its government with the institution of slavery.[1] And as it grew into a nation of masses—mass production, mass communication, mass democracy—there were plenty of worries about propaganda and mass delusion; as C. Wright Mills interpreted it, the nation evolved into a fully mass society rather than a community of publics.[2] Late nineteenth-century populists believed that the system was broken, much as today's do, and to travel back through the history of the notion of truth illustrates how today's discussions of deepfakes, photos suggesting mass crowds at Trump's inaugurations, and Pizzagate rumors fit into a longer discussion around the status of truth and rational governance.

While "fake news" goes back at least to the 1930s, the recent half century *did* place an especially high value on truth. Political anthropologist Kregg Hetherington notes that while *post-truth*, as a phrase, implies a recent past governed by truth—a questionable assumption—the post–Cold War era did in fact usher in "a series of premises about governance based on empirical knowledge," and he points to three keywords in particular that dominated development discourse in the 1990s: *transparency*, *information*, and *knowledge*.[3] Transparency was emphasized after the fall of the USSR, information rose upon the promise of the internet, and knowledge rose upon new hopes for the knowledge economy in the wake of the decline of manufacturing, Hetherington describes, suggesting that "this is what many have retroactively come to think of as 'Truth.'"

The eroded status of truth relates to both moral corruption and the rise of an algorithmically driven media ecology, among other things, but in this chapter, I'm concerned with the latter. Perhaps the resurgence of "fake news" is a blip as we adjust to the information age—the equivalent of a toddler acting out—before the age matures and before we've constructed systems of education and media consumption that are appropriate for an information age entwined with the Anthropocene. Or maybe fake news, troll farms, and bots are here to stay—in which case, rational discourse about and democratic governance of these technologies seems elusive.

Crucially, this moment is marked by a distrust not just of the numbers or the facts but of the authorities presenting them—a strain of antielitism that resonates throughout U.S. history but is particularly loud right now. As Diana Popescu writes, it is "established facts and agreed-upon truths" that post-truth groups deplore, and "insofar as post-truth is a new reality rather than an old but now more visible one, its novelty resides in distrusting established guarantors of truth, in part simply because they are established, while trusting grassroots observations over the venerable edifice of science."[4] This distrust of elites has consequences for both pro and counter geoengineering arguments. On one hand, we can imagine a Trump administration and base that goes forward with geoengineering despite the fact that many climate scientists are measured or cautious on climate engineering—or outright reject it. *The authorities of establishment science have simply got it wrong; this actually does work.* On the other hand (and I think this scenario is more likely), we can imagine an environmentalist response that argues that renewables can in fact be scaled up to the levels that would avoid climate damages or that soil carbon can save the day. *The authorities of establishment science have simply got it wrong; their assumptions about what is possible and their models are incorrect.* In short, either side could plausibly employ post-truth, antiestablishment thinking to guide a course of action in line with their culture and values. In what follows, I'll go into some of these arguments and their implications by considering first geoengineering research and then geoengineering implementation.

Geoengineering Research in Post-truth Conditions

The current media ecology has a few key features that impact how information about geoengineering circulates. First, a small number of people can claim a large part of the discursive sphere and amplify extreme positions. Second, news is narrowcast rather than broadcast, with polarization growing from repeated exposure to similar sources, and both domestic and foreign actors use this capacity. Third, the lack of face-to-face contact enables troll culture, and the tone of confrontation and hostility, with public and personal attacks on individuals doing their jobs, can inhibit public discourse just as much as misleading content would. Fourth, it seems increasingly evident that malicious state or nonstate actors can act in a coordinated fashion to influence discourse on a topic.

What do these media ecology features mean for climate engineering? First, public opinion on geoengineering could be easily influenced by someone with an agenda. It's possible to target and influence voters via social media—potentially, people who are vulnerable to messaging about geoengineering. The implications for the future evolution of technologies that can spread hyperbiased or false information—combined with firms who have data to conduct psychological targeting of voters—are real. Climate engineering could be dragged into this realm by either advocates or detractors, but either way, it will have a poor effect on public deliberation and could even influence state positions. These media ecology features may make geoengineering research less likely to be done, at least openly. Some readers will no doubt be thinking, "good." The problem, however, is that it's hard to imagine geoengineering research disappearing completely. Transparency is an important principle for geoengineering research, one emphasized by nearly every work on the subject. The rationale is that it would be better to have geoengineering research in the public sphere, where everyone can evaluate it and where people at universities can participate in it, rather than having it be led by other actors behind closed doors, such as the U.S. military, as suggested by Surprise.[5]

With a media ecology hostile to open discussion, though, the beliefs of the chemtrails conspiracy are in fact likely to self-fulfill: it's not hard to imagine a situation where geoengineering research does in fact go underground, because it has become so unpopular and researchers receive so many threats.

Serious thinkers have dismissed the chemtrails conspiracy, but it's significant for the politics of this issue. For those unfamiliar with it, the basic idea is that the government is already spraying chemicals out of airplanes for some purpose—ranging from mind control to, more recently, masking global warming. Once a fringe idea, a recent survey identified about 10 percent of Americans declaring the conspiracy as completely true and another 20–30 percent as somewhat true.[6] Interestingly, these views did not differ by party affiliation. The idea has comprised 60 percent of discourse about geoengineering on social media—that is, it has eclipsed the topic of climate engineering itself to become the dominant referent, and the study found Twitter to be a primary platform for this. One could be forgiven for assuming that people who believe in chemtrails are antiscience, but in fact, they are often very interested in taking environmental samples and measuring the chemicals they believe are being sprayed. The thing about chemtrails is that they can be seen with one's own eyes: the evidence is in the sky. As Tocqueville observed in *Democracy in America*, back in 1835, Americans "mistrust systems; they adhere closely to facts and study facts with their own senses."[7] Interestingly, belief in chemtrails is not a uniquely American phenomenon but a worldwide one, having spread from North America to the world via the internet.[8]

Again, rather than just a fringe conspiracy, this can be seen as a movement likely to have a political role. The chemtrails conspiracy is effectively a preexisting vehicle that any foreign or domestic actor could use to drive geoengineering discussion off the road. It appears to have arisen organically—Cairns traces it to the late 1990s[9]—though it may not continue to be a self-organized area of discussion. There are a few notable developments. First, there is a move from chemtrails toward an antigeoengineering movement, with some sites and actors now using the language of solar radiation management (SRM) science. This emergent movement seems to meet some of the basic criteria of being a social movement: dense networks, conflictual relations with opponents, training on how to attract supporters or win influence, and the use of both protest and legal strategies to challenge power holders. As a movement, it shares some concerns with other movements. Chemtrailers have real concerns about inequality, the degradation of the environment, and elite control of technology: "They look themselves in the mirror and see themselves as the upper well-educated 'scientists,' far above all below them"; . . . "The Zillionnaires who are richer than imaginable want the entire world to operate on an algorithm."[10] We can expect these underlying tensions around inequality to grow under the current administration, thus potentially attracting new people to an antigeoengineering movement. It's possible that forming an opinion on geoengineering becomes not an impersonal assessment of a technology but an identity position.

Many people who subscribe to chemtrail beliefs are concerned about the ways in which our society is ill; as Balalaki reads it, the chemtrail narrative is a story of loss. Moreover, what is happening to the biosphere can rightfully produce collective trauma: anyone paying attention to both the extinction of species and the failure of leaders to deal with or properly acknowledge this crisis should be both outraged and sad. As long as there are not open and recognized mainstream fora in which to work through the unraveling tragedy, movements like this one—or antivaccination, or skepticism about electromagnetic fields, or Morgellons nanofibers, or other arenas where technology and environment intersect—will probably continue to bubble up. Scientists would do well to communicate their work in a way that acknowledges some of their community's underlying concerns: about inequality of income and opportunity, about extinction and separation from nature. The frame of geoengineering needs to be set so that it can acknowledge and not occlude the human and nonhuman pain of these times. This will likely go contrary to the specialization and narrow foci that define our academic comfort zones.

All in all, it is easy to be pessimistic about post-truth effects on solar geoengineering research. What about post-truth effects on carbon removal research? Likely, they will depend on what ideology a person already has. It's common to link post-truth with right-wing populism and a right-wing populism that is hostile to climate change science.[11] In this view, climate policy is a tool of elites to enact an agenda of regulation, taxation, and control, and carbon removal could easily be viewed as an extension of that.

But it's not the case that right-wing populists are necessarily post-truth, while liberals are the guardians of truth and facts. Liberals might find that certain forms of carbon removal fit into their value systems—such as soil carbon sequestration—and inflate facts or numbers about how much it can do to mitigate climate risks. Anecdotally, I went to one conference keynote lecture where it was stated that soils could sequester twice as much carbon as we are currently emitting, an estimate that stretches beyond optimism to the point of departing reality entirely. It's possible that post-truth interpretations of some carbon removal solutions could really spur research into them. The same goes for calculations of the potential for carbon capture and utilization, often referred to as being a potential $1 trillion[12] market by 2030, without any evidence—though business hype has of course been post-truth before post-truth was even a phrase.[13]

On the other hand, if carbon removal is seen as a threat to mitigation, there will be intense discussions around the potential of renewables to manage climate risks, with claims that may edge into post-truth territory—again, because these facts about how quickly renewables can be scaled fit with one's values. Meanwhile, people far away from these Twitterverses and conference buffets will be actually losing their crops, health, and homes from climate change.

In any case, when it comes to research on carbon removal, two things seem clear. First, people will likely oppose carbon removal on principle if it comes from the wrong messenger, even if they might like it in other circumstances.

Second, if carbon removal is authored as a liberal carbon management scheme coming out of think tanks in Washington or tech groups in Silicon Valley, this could make it even less likely to be a source of technological progress that is grounded in rural economies. The success or failure of carbon removal is a prospect that is absolutely relevant for solar geoengineering too because ceasing solar geoengineering would hinge on the development of massive carbon removal.

Geoengineering Implementation in Post-truth Conditions

I've suggested that a post-truth media ecology dims the prospects of transparent solar geoengineering research and could either hamper or spur on research of various carbon removal techniques, depending on who takes leadership of the issue and whom it is associated with.

When it comes to geoengineering implementation, however, the picture gets even worse because of the link between post-truth and authoritarianism, which can be viewed as related but distinct phenomena. *Post-truth* I use as a term describing the media ecology that is used effectively by those with authoritarian leanings. As Stewart Lockie writes, "What we have seen in so-called post-truth politicians is an authoritarian impulse that promises to be both reckless and destructive—an impulse all too comfortable with the deployment of propaganda, vilification and intimidation."[14] Autocrats are thriving in this media ecology, even as it promises to offer a voice to the people. My main concern regarding geoengineering implementation, then, is how post-truth conditions make it more likely for the implementation to be autocratic. Let's consider three scenarios for this.

The first scenario is easy to imagine and has been oft-discussed: a post-truth society leads people to begin solar geoengineering without really understanding the risks involved, and/or it leads to mitigation deterrence where solar geoengineering is misleadingly viewed as a substitute for mitigation. In either case, the immediate rulers of the status quo stand to benefit in the near term.

Now consider a more specific scenario: solar geoengineering enables regime continuation and thus actually increases autocracy. Solar geoengineering incentivizes regime continuation on a very fundamental level because of the threat of ecological disruption from a termination shock. It involves basically banking on the notion that a stable regime of one form or another will exist to continue the solar geoengineering for decades. In a "weak" form of this, ruling elites might suggest that they are doing a good job, and thus term limits should be removed so that they can continue benevolently watching over and executing the geoengineering program. In a "strong" form of this, ruling elites use the threat of geoengineering disruption and subsequent termination shock to quell dissent and basically do whatever they want—they are indispensable for climatic maintenance and so can get away with murder.

Finally, let's leave aside solar geoengineering and consider how carbon removal might be implemented to further autocracy. Consider platform dominance: one platform becomes the key way to organize trading of carbon, compensating for residual emissions, monitoring removal and geological disposal, and the like. It's not hard to imagine this, because platforms have network effects that lead to the dominance of the few. Now imagine that the platform is set up to benefit a corrupt regime; one who is skimming off the top somehow, siphoning transaction fees or worse. No one wants to risk limited progress in removing carbon, so they tolerate it. The regime grows up with the carbon removal platform and sets it up in ways that benefit itself.

These scenarios are of course speculative, but disturbingly, neither are they impossible to imagine. They may be low-probability, high-risk outcomes, and so it's worth thinking about how to shape these technologies to prevent such outcomes. I think it is important to remain agnostic on the matter of choice, of intention, and of the possibility of shaping outcomes here. Going back to the history told in *These Truths*, Lepore centers the story on a question posed by Alexander Hamilton in 1787 and that she seems to consider still open: whether good government can be established through reflection and choice, or whether men are forever destined to be governed by accident and force.[15] This basic question relates to climate engineering governance too: Can climate engineering governance be established through choice, reflection, and science—or will it be rather a matter of accident and force? Ironically, closing down on the latter assumption places one in the absolute cynicism of the chemtrailers, caught inside the machinations of violent powers.

Fixing Things

The remedies for post-truth and authoritarian tendencies aren't quick or simple, and by and large, they aren't technical. They include at least three long-term pillars: income equality, electoral reform, and education. These pillars would hold up a saner version of climate policy, and climate engineering governance should also start from this point.

Consider inequality, which is often bounded off in some social realm distant from the environmental or technical issue of climate engineering. One could imagine policy around carbon removal, and agriculture more broadly, being designed in ways that bring good infrastructure and opportunities to rural regions—contrasted with, for example, heavily financialized policy for pricing carbon that makes some bankers richer by selling complex financial products and doesn't accomplish much else. Electoral reform has implications for climate engineering governance as well. People who aren't even allowed to vote or who live in districts who have been gerrymandered into a situation where their vote isn't worth that much are going to feel disenfranchised from civic life more generally, which impacts their inclusion on decision-making about geoengineering. Advocates of social justice are going to be working on these things anyway for reasons far beyond geoengineering—the point is simply to say that discussions of geoengineering should recognize these problems as fundamental to geoengineering politics, not separate from them.

We also need to treat our media ecology.

There actually are some technological fixes here, and we need to hold tech firms accountable for their algorithms. In the wake of endless scandals around big tech and social media, there is no shortage of good writing on how to regulate and remedy the internet, from think tanks such as Data & Society to the pages of established journals and outlets like *Science* and the *Economist*. Legislation toward decentralization, or treating tech titans like utilities, are frequently called-for options, though getting serious about regulating tech firms will require a new way of thinking about them. Communications scholar Fred Turner notes that "one of the deepest ironies of our current situation is that the modes of communication that enable today's authoritarians were first dreamed up to defeat them"; his essay illustrates how the right wing took advantage of a liberal effort to decentralize the media.[16] Crucially, he points out that the new authoritarianism represented by the likes of Trump and the white supremacist Richard Spencer isn't just a product of media ownership but "a product of the political vision that helped drive the creation of social media in the first place—a vision that distrusts public ownership and the political process while celebrating engineering as an alternative for of governance." (Climate engineering is a related fantasy—that engineering will be cleaner than politics, that it allows some kind of escape from the messiness.) Alongside antitrust and other regulation, I would add that the real solutions include entangled long-term issues of science/information literacy and mental health.

Indeed, there's another part that's less obvious to the liberal mind: making a place for emotions in discourse about climate engineering and other topics. The understanding of post-truth politics, writes Rhys Crilley, is flawed in that it tells a story about the victory of emotion over reason—but emotions were always there in politics and society. "Feminists, critical theorists and others outside the mainstream of academic inquiry have argued so for decades," Crilley writes; what's new is "the recognition, both within the study of these respective fields and within wider public discourse, that emotions matter."[17] The remedy for post-truth effects isn't to further denounce emotional thinking but to bring it in. Jasanoff and Simmet also suggest deconstructing binaries like emotion/reason; they talk about true/false or science/antiscience and how these binaries work to strengthen political polarization. Instead, they call for a discourse that understands that truth in the public sphere doesn't exist as some independent set of facts to be discovered: "We need above all to resist the unthinking reduction of lived realities to technical facts, assuming that both are singular and both can be ascertained through apolitical delegation to scientific consensus."[18]

These general insights have deep value for the specific issue of geoengineering. The immediate tendency, in this chaotic post-truth media ecology, is to hold up what's actually true. *Jet planes make contrails. The world is warming.* What we actually need to do seems counterintuitive: simultaneously explain what's empirically true while also engaging with the emotional parts of this debate. *We are living a tragedy. It's filled with injustice. So much has been lost. Our leaders failed us; our system failed us. We are going into the unknown. There are still actions we can take to make things better.* Only then can we look more honestly at our prospects and hope to avoid the worst-case failures of these technologies.

Notes

1   Jill Lepore, *These Truths: A History of the United States* (New York: W. W. Norton, 2018), 191.
2   Lepore, 566.
3   Kregg Hetherington, "What Came before Post-truth?," *EASST Review* 36, no. 2 (2017).
4   Diana Popescu, "What We Talk about When We Talk about Post-truth," *Aeon Magazine*, April 21, 2018, http://aeon.co/ideas/what-we-talk-about-when-we-talk-about-post-truth.
5   Surprise, this volume, chap. 13.
6   Dustin Tingley and Gernot Wagner, "Solar Geoengineering and the Chemtrails Conspiracy on Social Media," *Palgrave Communications* 3, no. 12 (2017).
7   Lepore, *These Truths*, 211.
8   Alexandra Bakalaki, "Chemtrails, Crisis and Loss in an Interconnected World," *Visual Anthropology Reivew* 32, no. 1 (2016): 12–23.
9   Rose Cairns, "Climates of Suspicion: 'Chemtrail' Conspiracy Narratives and the International Politics of Geoengineering," *Geographical Journal* 182, no. 1 (2016): 70–84.
10 Comments on Geoengineering Watch (website), geoengineeringwatch.org.
11 See Matthew Lockwood, "Right-Wing Populism and the Climate Change Agenda: Exploring the Linkages," *Environmental Politics* 27, no. 4 (2018).
12 The symbol $ refers to U.S. dollars.
13 The Global CO_2 Initiative (website), http://www.globalco2initiative.org.
14 Steward Lockie, "Post-truth Politics and the Social Sciences," *Environmental Sociology* 3, no. 1 (2017): 1–5.
15 Lepore, *These Truths*.
16 Fred Turner, "Machine Politics: The Rise of the Internet and a New Age of Authoritarianism," *Harper's*, January 2019.
17 Rhys Crilley, "International Relations in the Age of 'Post-truth' Politics," *International Affairs* 94, no. 2 (2018): 417–425.
18 Sheila Jasanoff and Hilton R. Simmet, "No Funeral Bells: Public Reason in a 'Post-truth' Age," *Social Studies of Science* 47, no. 5 (2017): 751–770.

Acknowledgments

J. P. Sapinski would like to thank the Corporate Mapping Project, the Social Sciences and Humanities Research Council of Canada, and the Université de Moncton Faculty of Graduate Studies and Research for their financial support at various stages of this project. Holly Jean Buck and Andreas Malm gratefully acknowledge funding from the Swedish research council Formas, grant Dnr 2018-01686, which contributed to research that went into this book. All our thanks to Kevin White and Karine Godin at Université de Moncton for their assistance with referencing and indexing.

Notes on Contributors

HOLLY JEAN BUCK is an assistant professor of environment and sustainability at the University at Buffalo. Her research interests include agroecology and climate-smart agriculture, energy landscapes, land use change, new media, and science and technology studies. She is the author of *After Geoengineering: Climate Tragedy, Repair, and Restoration* (2019). She has written on several aspects of climate engineering, including humanitarian and development approaches to geoengineering, gender considerations, and the social implications of scaling up negative emissions.

WIM CARTON is an assistant professor at the Lund University Centre for Sustainability Studies, Sweden. His main research interest is in the political ecology and political economy of climate change mitigation. Wim has previously worked on market-based mechanisms for climate mitigation, notably emissions trading and forest-based carbon offsetting in the Global South. He is currently the cocoordinator of a four-year research project on carbon dioxide removal and the politics of integrated assessment modeling.

LAURENCE L. DELINA conducts research at the Frederick S. Pardee Center for the Study of the Longer-Range Future at Boston University. Aside from publishing on climate change, energy, and sustainable development in high-impact journals, he has also authored four books on rapid climate mitigation, accelerating energy transitions, and emancipatory and transformative climate actions. He has consulted for the United Nations and Oxfam, was a Rachel Carson fellow, and was a Visiting fellow at Harvard Kennedy School.

LILI FUHR leads the International Environmental Policy Division of the Heinrich Böll Foundation's head office in Berlin and has a background in geography, political science, sociology, and African Studies. She focuses on issues of international climate and resource politics. She has been a board member of the ETC Group since 2015 and a founding board member of the Climate Justice Fund since 2017. She is a coauthor of *Inside the Green Economy—Promises and Pitfalls* (2016) and blogs at http://www.klima-der-gerechtigkeit.de (in German).

MADS DAHL GJEFSEN is a senior researcher at the Work Research Institute (AFI), OsloMet—Oslo Metropolitan University, Norway. His background includes a PhD in science and technology studies (STS; University of Oslo, 2015) as well as project management for the Norwegian Board of Technology and in a public university technology transfer organization. He is broadly interested in the societal orders reflected in envisioned technological futures and in the political and commercial affordances of technological promise making.

ANDREAS MALM teaches human ecology at Lund University. He is the author of *Fossil Capital: The Rise of Steam Power and the Roots of Global Warming* (2016) and *The Progress of This Storm: Nature and Society in a Warming World* (2018).

NILS MARKUSSON is a lecturer at the Lancaster Environment Centre (LEC), Lancaster University. He has a wide-ranging background in engineering, innovation policy, innovation studies, and STS and most recently cultural political economy (CPE). He has previously researched carbon capture and storage (CCS), focusing on innovation processes, learning, and lock-in. He also published on the social construction of climate engineering. His current work is about possible interaction effects between greenhouse gas removal technologies and climate mitigation efforts.

DUNCAN McLAREN is a visiting researcher at Linköping University, Sweden, and professor in practice and research fellow at Lancaster University, where he examines the interactions between greenhouse gas removal and emissions reduction in climate policy. His PhD, completed in 2017, examined the justice implications of geoengineering. In his previous career, Duncan worked as an environmental researcher and campaigner, most recently as the chief executive of Friends of the Earth Scotland from 2003 to 2011.

INA MÖLLER has a PhD in political science and master's degree in environmental studies. She conducted her doctoral research at Lund University in Sweden, where she studied the emergence of geoengineering on the global political agenda. Her research has been published in the journals *Environmental Politics* and *International Environmental Agreements*, and she has acted as an adviser to the German Agency for the Environment on the subject of geoengineering governance.

CHRISTIAN PARENTI is an associate professor of economics at John Jay College, City University of New York. His teaching and research focus on social justice, environmental justice, globalization, climate change and climate justice, sustainable energy, American economic and environmental history, and political violence. He has reported extensively from Afghanistan, Iraq, and various parts of Africa, Asia, and Latin America for the *Nation*, *Fortune*, the *London Review Books*, the *New York Times*, and other publications.

ANNE PASEK researches climate communication and the changing cultural politics of carbon. She is an assistant professor of cultural studies and environment at the University of Trent. Her work is interdisciplinary, drawing on the traditions of new materialisms and feminist STS as well as culture, queer, media, and communication studies. She is the reviews editor of the *Journal of Environmental Media* and has been published in journals such as *Culture Machine*, *Feminist Media Studies*, and *Photography and Culture*. She is a profligate knitter, a strong proponent of low-carbon research methods, and an occasional labor organizer.

J. P. SAPINSKI is an assistant professor of environmental studies and public policy at Université de Moncton in Canada. His work draws from the critical political economy and power structure research traditions to map out the constellations of corporate interests involved in the politics of climate change and energy, including geoengineering politics. He is a coauthor, with William K. Carroll, of *Organizing the 1%: How Corporate Power Works* (2018).

LINDA SCHNEIDER is the senior program officer for international climate policy at the Heinrich Böll Foundation's head office in Berlin and holds MA degrees in international relations from the Free University of Berlin, Humboldt University Berlin, and the University of Potsdam as well as a BA in political science from Hamburg University. Linda is active in the climate justice movement in Germany. She is also a member of the German chapter of ICAN—the International Campaign to Abolish Nuclear Weapons.

JENNIE C. STEPHENS is the director of Northeastern University's School of Public Policy and Urban Affairs and dean's professor of sustainability science and policy. Her research, teaching, and community engagement focus on socialpolitical aspects of renewable energy transformation, energy democracy, climate resilience, reducing fossil fuel reliance, gender diversity, and social, economic, and racial justice in climate and energy policy. She earned her PhD and MS at the California Institute of Technology in Environmental Science and Engineering and her BA at Harvard in Environmental Science and Public Policy.

KEVIN SURPRISE holds a PhD in geography from Clark University and is currently a visiting lecturer in environmental studies at Mount Holyoke College. His research explores the political economy of global climate change, with a focus on the development of solar geoengineering technologies. This research situates solar geoengineering as an emergent imperial tactic, examining the relationships between geoengineering, ecological crises, and spatiotemporal fixes; ideologies and sources of funding shaping solar geoengineering research;

and the ways in which the U.S. military and security community is influencing solar geoengineering development.

TINA SIKKA is a lecturer in media and culture at Newcastle University in the United Kingdom. Her current research focuses on the feminist and intersectional study of science and technology applied to environmental/climate science, health, and food. Her most recent book is titled *Climate Technology, Gender, and Justice: The Standpoint of the Vulnerable* (2018).

DAVID TYFIELD is a reader in environmental innovation and sociology at the Lancaster Environment Centre (LEC), Lancaster University. He is executive director of the Joint Institute for the Environment (JIE), Guangzhou, China, and codirector of Lancaster's Centre for Mobilities Research (CeMoRe). His research focuses on the interaction of political economy, social change, and developments in science, technology, and innovation, with a particular focus on issues of low-carbon transition in China, especially urban e-mobility, which he has been studying since 2007. He is the author of *Liberalism 2.0 and the Rise of China* (2018) and is a coeditor of *Mobilities* journal.

KYLE POWYS WHYTE holds the Timnick chair in the humanities in the Department of Philosophy at Michigan State University and is a faculty member of the environmental philosophy and ethics graduate concentration. His primary research addresses moral and political issues concerning climate policy and Indigenous peoples and the ethics of cooperative relationships between Indigenous peoples and climate science organizations. He is an enrolled member of the Citizen Potawatomi Nation in Shawnee, Oklahoma. His articles have appeared in journals such as *Climatic Change, Environmental Justice, Hypatia, Ecological Processes, Synthese, Human Ecology, Journal of Global Ethics, American Journal of Bioethics, Journal of Agricultural & Environmental Ethics, Ethics, Policy & Environment*, and *Ethics & the Environment*.

RICHARD YORK is a professor of sociology and environmental studies at the University of Oregon. His research focuses on environmental sociology, ecological economics, animal studies, and the sociology of science. He has over one hundred publications and has received numerous awards for his scholarship and teaching.

Index